Power Systems

Electrical power has been the technological foundation of industrial societies for many years. Although the systems designed to provide and apply electrical energy have reached a high degree of maturity, unforeseen problems are constantly encountered, necessitating the design of more efficient and reliable systems based on novel technologies. The book series Power Systems is aimed at providing detailed, accurate and sound technical information about these new developments in electrical power engineering. It includes topics on power generation, storage and transmission as well as electrical machines. The monographs and advanced textbooks in this series address researchers, lecturers, industrial engineers and senior students in electrical engineering.

** Power Systems is indexed in Scopus**

More information about this series at http://www.springer.com/series/4622

Viktor Józsa · Róbert Kovács

Solving Problems in Thermal Engineering

A Toolbox for Engineers

 Springer

Viktor Józsa
Department of Energy Engineering
Faculty of Mechanical Engineering
Budapest University of Technology
and Economics
Budapest, Hungary

Róbert Kovács
Department of Energy Engineering
Faculty of Mechanical Engineering
Budapest University of Technology
and Economics
Budapest, Hungary

ISSN 1612-1287   ISSN 1860-4676   (electronic)
Power Systems
ISBN 978-3-030-33477-2   ISBN 978-3-030-33475-8   (eBook)
https://doi.org/10.1007/978-3-030-33475-8

This Springer imprint is published by the registered company Springer Nature Switzerland AG
The registered company address is: Gewerbestrasse 11, 6330 Cham, Switzerland

to the memory of our young colleague,
Gergely Novotni
and
to my wife, Ági, and our daughter, Julianna
Viktor Józsa

to my beloved parents
Róbert Kovács

Preface

Thermal analysis is often omitted in practical research and industrial applications due to its secondary importance, hoping that heat will find its way in or out while the equipment is fully functional. The market pull is cruel, there are strict deadlines which usually does not allow a few extra days and engineer hours to spend. Therefore, thermal engineers are rarely employed in small companies who actually have serious thermal problems, leading to significant losses. In most cases, this viewpoint is initially viable and safe. The reliable operation will be guaranteed if enough tests were performed with the equipment. Nevertheless, malfunctions, crashes, and other events are often originated from thermal problems, and they come up after weeks, months, or years of use. The principal aim of the present book is providing practical tools for solving thermal problems at all levels from the beginner to the researcher. It is always better to have a rough estimate in a paper (yes, even today, papers help a lot in solving complex problems to have a good overview from a distance of our eyes) than entirely omitting thermal evaluations since one does not have the background or confidence in using the state-of-the-art calculation tools.

Solving thermal problems require abstract thinking since a small temperature difference does not hurt during testing, does not have color, odor, direction, and others. Furthermore, the time scale of thermal problems is usually exceeded that of chemical, fluid dynamical, and mechanical problems of the same physical size by orders of magnitude. Therefore, old churches and castles with thick walls vary their inside temperature marginally over the year while a tree trunk floating on a river may travel kilometers in an hour. Instead of digging deep into the modern methods which might be obsolete in few years or a decade, the main focus of this book is on discussing the necessary viewpoint to solve thermal problems, emphasizing the state-of-the-art challenges what should be solved for reliable operation or estimation, and modernly, ensure smart energy use. This latter criterion became standard since the beginning of the century; however, most of our current buildings, machines were designed earlier when these requirements were secondary, calling for a different design philosophy.

This book is divided into six chapters which are loosely connected but represent different tools for solving thermal problems. Chapter 1 is focusing on the general approach of thermal problems, which is recommended to start with for beginners and industrial engineers. It discusses the fundamental concepts of a thermal system, including boundaries, and basic calculation methods for estimating the thermal behavior.

Chapter 2 is the summary of the governing equations of heat conduction to provide a framework of advanced modeling. It gives a brief overview of the mathematical structure of particular models, including heat conduction, fluid mechanics, diffusion and mechanics, situating them onto the same thermodynamic ground. It mostly uses the approach of Classical Irreversible Thermodynamics but also presents its possible extensions.

Chapter 3 is focusing on thermal problems in energy engineering through several examples, concentrating on renewable technologies. To minimize our ecological footprint, we need efficient machines for a given purpose. However, the losses are often realized as heat transfer or dissipation to the ambient. As our present machines which we use today will likely remain in service for decades, their retrofit for efficient use of thermal energy is the problem of the present to solve. Hence, besides solar and wind applications, combustion is also discussed regardless that is unpopular in the media.

The space industry has exploded since the introduction of private companies in the past decade with viable plans to make commercial use of space resources. Hence, Chap. 4 is dedicated to discussing the thermal balance and its challenges in space. Nevertheless, most of the content can be directly applied to vacuum applications on Earth as well.

Chapter 5 focuses on the experimental proofs of non-Fourier thermal conduction. That is, it includes low and room temperature measurements in which various phenomena are observed, such as second sound and ballistic propagation. It was initially measured in cryogenic conditions, which is usually far from applications. Nevertheless, the non-Fourier behavior was proven to be present even in room temperature under unsteady conditions, requiring heterogeneous materials or nano-sized objects. Typically, composites and layered structures from distinct materials fall into this category. Since the appearance of the non-Fourier behavior depends on the particular length and time scales, one should be aware of these, and that advanced thermal conduction models exist which are likely to appear in commercial simulation software codes in the near future.

Numerical methods are discussed in Chap. 6, focusing on the implementation of boundary conditions both from numerical and analytical aspects. Furthermore, a particular way of error estimation is demonstrated on a conservative system which makes apparent how easy to obtain stable, but unphysical solutions. It affects all software. The relevance of this chapter is that simulation modules are available in a rapidly increasing number of engineering software. It is often a few clicks on a 3D drawing to get a thermal analysis with a convergent result. However, the user must be aware that the model always remains an estimate of the real process. Therefore, careful validation and method analysis are required prior to making any decision based on a colorful result to avoid spectacular mechanical, thermal, and financial failures.

The problems which we feel in our skin range from the size of our planet down to the nano-size scale in the semiconductor industry where the continuum-based physics is still applicable. Consequently, understanding the governing logic and using the appropriate mathematical tools allow us to build a versatile knowledge independent of the given application and available software in the market.

Wishing you joy and persistence in your life and career,

Budapest, Hungary   Viktor Józsa
                                                                                 Róbert Kovács

Acknowledgements

The authors thank Tamás Fülöp, Péter Ván, Mátáys Szücs, Gregory Kowalski, and the BME Combustion Research Group for their valuable ideas, and their support.

The work was supported by the grants National Research, Development and Innovation Office—NKFIH 116197(116375), 124366(124508), 123815, KH130378, FK124704, and FIEK-16-1-2016-0007, New National Excellence Program of the Ministry of Human Capacities ÚNKP-19-4-BME-213, the János Bolyai Research Scholarship of the Hungarian Academy of Sciences, and the NVIDIA Corporation with the donation of the Quadro P6000 which was used for the calculations presented in Chap. 4. The research reported in this paper was supported by the Higher Education Excellence Program of the Ministry of Human Capacities in the frame of Nanotechnology research area of Budapest University of Technology and Economics (BME FIKP-NANO).

We would like to thank the Elsevier, APS, and Taylor&Francis for the permission to reuse the relevant figures.

Contents

Chapter 1
The Way of Problem Solving in Thermal Engineering

1 The Toolbox

Everyone wants to solve the problems in less time and be more productive. This is especially true for engineers who always keen to find the next challenge and having low motivation when they have to return to the same problem multiple times. Consequently, the only solution to minimize the overall effort is following a precise step-by-step procedure and making notes and documentation of the solved problem. This is the backbone of clear and effective work. To detail this way of thinking, a proposed framework is discussed in the first section. As it is practice-oriented, the content is supported by tips and practical explanations instead of literature references. The last section discusses data uncertainty. To evaluate a measurement, it is mandatory to know the uncertainty of the used devices, i.e., they have to be calibrated. Since the uncertainty of a derived quantity is highly important in model validation, the calculation procedure is briefly discussed, also taking correlated quantities into account. Hence, this knowledge is useful for both experimental and simulation engineers.

2 The Steps of Problem Solving

When one launches a numerical program to solve a thermal problem, a lot has been decided before with good reason. It is the determination of the thermal system and its connection to the ambient, i.e., the general behavior of the boundary conditions. To get rid of the limited capabilities and streamlined procedures of computer codes, grab a piece of paper, and start freely. The steps are the following, using the list of Struchtrup [1], and extending the points with personal experience on solving various industrial problems.

1. **List the problems to be solved!** If the questions are not crystal clear at the beginning, it is possible that another evaluation or a completely new model or approach will be necessary later on, after the contract has been signed. In addition,

© Springer Nature Switzerland AG 2020
V. Józsa and R. Kovács, *Solving Problems in Thermal Engineering*, Power Systems,
https://doi.org/10.1007/978-3-030-33475-8_1

list what will not be covered in the work. Make notes of meetings and share with the partners to let everyone know what is exactly the problem to be solved and how the work progresses.

2. **Know well the purpose of your system.** Theory often needs to be supplemented by practical information about the habits of system use. Hence, arrange a few hours to talk with the operators to understand their problems. The leader, who asks your help, usually know a little about them. To answer not only the asked but the hidden questions as well for overwhelming success, you have to know both the theory-related and the operation-originated issues.

3. **System identification.** If the system is too complex to handle at once, you can break it down to subsystems. This is the first step in problem solving. Due to its highlighted role, Sect. 2.1 details it through an example of a solar collector.

4. **Note the limitations.** Example questions: Is there a temperature which is not allowed since the used material will melt? What is the maximum allowed pressure? Is the selected material weatherproof (for outdoor applications)? Is it possible that small animals will damage the equipment? For instance, birds love building nests from thermal blankets. This cannot be calculated from any differential equation.

5. **Summarize the variables to be calculated.** Compare their number to the number of equations for checking the match. This is a typical step which is often overlooked. The partner doesn't understand the problem and doesn't provide enough details, but it is expected that you can calculate it in a commercial numerical code because it is expensive. This is the point where you have to stand up and show the problem solving process up to this point, and show what else is needed to make the problem solvable.

6. **Prepare the framework.** Draw the geometry, define the boundary conditions and the set of equations to solve. Chapters 2 and 6 provide deep details on this procedure to summarize the mathematical model. However, it is advised to do all the steps in a paper/spreadsheet at first, only then turn to numerical software if necessary. Add the physical effects step-by-step, e.g., start with thermal conduction and convection at first in a steady case, only then add radiation and perform unsteady calculations. Brief details on models are summarized in Sect. 2.2.

7. **Perform the calculations.** If the model is complex, always go step-by-step to allow checks that everything is working well. It is always hard to find the error when a complex physical behavior is analyzed. If the physical model is working, only then scale the problem up, i.e., use large mesh size on a server, utilizing hundreds/thousands of cores.

8. **Evaluation.** Evaluate the whole system and its parts. If complex simulations were performed, check their match with an analytical estimation. Do the results make sense? Is the system work properly? If the task is a design, optimize the system performance to minimize the losses.

9. **Is it the best solution?** Is the analyzed system appropriate for the task? Are there better solutions on the market which are comparable in price while having a better performance?

+1 **Write comments to remind yourself.** It is rather common that you need to revisit your calculations/software code after a week, a month, or a few years. These notes will be invaluable at such times.

2.1   The Thermal System

According to the third step of problem solving, system identification, the boundaries of the thermal system has to be determined, considering thermal, contacts to the ambient. The universe—with unknown boundaries—is too large and complex to make simulations on it. Therefore, a reasonable-sized control volume should be considered initially which encompasses the whole system to be analyzed. The control volume may vary its shape, size, or move over time if it makes the calculations easier. Always perform the possible simplifications since not only the calculation time will be lower either on paper or on a computer, but error searching will also benefit from this.

After these 'how to' advices, the general considerations are qualitatively shown through an example of a solar collector, shown in Fig. 1. Note that only key heat transfer processes are presented here. For a complete list with a solution by using a thermal network method, see Sect. 2.3 of Chap. 3.

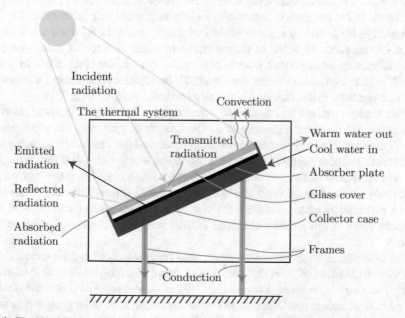

Fig. 1 The identification of the thermal system in the case of a solar collector

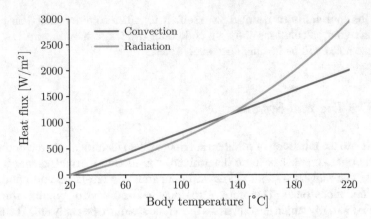

Fig. 2 Comparison of the heat flux by thermal convection and radiation. $T_{ambient} = 20\,°C$, heat transfer coefficient is 10 W/(m²·K), and the view factor and the emissivity are both unity

Initially, the thermal contacts between the ambient and the system need to be determined. In most cases, it is the appropriate consideration of thermal conductivity, heat transfer, and thermal radiation. The theory and the equations for the first two processes can be found in Chap. 2, while thermal radiation is discussed in Chap. 4.

In the present case, thermal radiation is the dominant heat transfer way how a solar collector receives heat. Even though the radiation from the Sun to the surface of the collector is obvious, radiation plays a non-negligible role in heat transfer. It is illustrated in Fig. 2 for a sample condition which can occur in the open atmosphere, therefore, automatic rejection of thermal radiation, referring to the 'small' temperature differences, may cause a notable bias in the results. Note that radiation scales with T^4 while convection scales only with T. In addition, convection is assumed here as a constant value, however, it usually increases slightly with T.

Heat transfer should be considered everywhere when a solid is surrounded by a fluid that can be either a gaseous or a liquid medium. Heat transfer coefficient is usually estimated by using experiment-based literature data. The most comprehensive book is referred here to provide a solid starting point, which is the VDI Heat Atlas [2].

The calculation of thermal conduction is the easiest from the three heat propagation ways if the thermal conductivity of the material is fully known, and the case can be assumed as steady. Under these conditions, Fourier's law can be used. For unsteady cases at small time scales, non-Fourier analysis is necessary, discussed in Chap. 5.

The boundaries of the thermal system also facilitate checking the calculations. The thermal balance of inlet and outlet heat flows should give zero if the case is steady and there is no inner sink or source. If the system is divided into subsystems or a numerical code is used, the contacts between bodies have to be analyzed. Instead of automatically neglecting certain thermal contacts, calculate them first and make notes on it to know its extent in the heat balance compared to other terms. When a manual procedure is chosen, the direction of the heat flow has to be assumed first, and the sign of the result will automatically correct it if necessary. Instead of guessing the actual heat flow direction, be consequent and use the same convention for all parts when calculating heat transport since if heat is transferred from one body to another, the sum has to be zero.

2.2 Thermal Models

All practical devices feature several elements which were designed to provide mechanical support, make manufacturing and assembly easier, ensure safer operation, or reduce costs. Hence, it is the first task to analyze the geometry and remove the thermally irrelevant parts. There is no general recipe since bolts and nuts are often omitted, however, when a thermal blanket or a shield is fixed with such a part, the heat transport through it will be considerable, hence, omitting it here will surely lead to biased results. Instead of promptly jumping into model simplification, it is advised to perform overall calculations to understand the system better, following the first item of the below list.

- Concentrated parameter modeling. It can be solved in a paper or by using a spreadsheet in more complex systems. The result is an overall energy balance which can be used for reference, preliminary design, or concept evaluation. For a faster solution, usually steady conditions are considered. Since this model requires the least effort, it is highly advised to perform this analysis—the higher the complexity, the higher the value of these calculations.
- Thermal network analysis. By using the key bodies, heat transport between each part and their ambient is calculated, similarly to the free body diagram in mechanics [3–5]. Since the subcomponents are concentrated parameter models, its concept is similar to the simplest variant discussed above where the calculations are performed for only one body. The solution of this system most probably requires a computer program which can be either a commercial code or a self-written one. The set of equations to solve can easily exceed the quantity which can be handled in paper and may require inner iteration cycles. The advantage of such models is that up to a few tens of bodies is that the solution takes typically a few seconds. This approach is excellent for design optimization [6–8] where a few thousand runs might be necessary. Another reason is parameter identification [9–11] for an advanced control system where online operation and actuation is desired [12–14]. If the product to be checked exists, the design is complete, or the detailed behavior of the system is interesting, not its overall characteristics, this type of analysis is omitted in favor of advanced numerical methods which are discussed next.
- Finite element/volume method (FEM/FVM) analysis represent the state-of-the-art in engineering problem solving. The result shows the detailed behavior in an extent which is usually not accessible by measurements. The software codes are highly generalized, and the use is streamlined. The leading programs feature external connection capabilities which allow scripting and data transfer between

various software. The commercial software usually comes with technical support and have a good theory and user guides while it takes significantly more time to learn open source systems as they typically don't feature these; mostly forums exist. Before choosing FEM/FVM software at a corporation, financial analysis should be performed. Beyond the price of the license, the salary of highly qualified engineers who can appropriately operate it, and the price of a computational infrastructure must be available as well. When two companies compete for a project with one using commercial codes and the other using open-source software, the difference in the offer is mostly $\sim$10−20% in favor of the open source variant. Nevertheless, equivalent solutions can be obtained by using either commercial or open-source software [15, 16].

Either model is about to be applied from the list above, it will be a simplification of reality and require validation, using measurement data of the given or a similar device. The uncertainty, detailed in Sect. 3, should always be considered. When a calculation is complete, the question is the correlation between the model and the real operation. As a last tip for thermal calculations do not afraid of using simple techniques instead of complex software which provide colorful results. Error identification and recalculation always require less effort in the case of simple models. Nevertheless, when the problem really demands FEM/FVM, use it with confidence as a part of the above-mentioned list of problem solving.

3   Data Uncertainty

In thermal engineering, the goal is to predict or reproduce a thermal behavior of a device, material, or a structure. Therefore, the set of equations necessarily contain material properties (e.g., thermal conductivity, kinematic viscosity, and specific heat) and state variables (e.g., pressure, temperature, and enthalpy). Even if the physical models would be 100% accurate, the measured quantities which are used for the calculation always burdened by uncertainty, therefore, the result cannot be more certain than that. Through an example of a turbocharger, uncertainty determination is presented by Olmeda et al. [17].

Carstens et al. [18] provide a comprehensive review of the uncertainty of various sensors used in energy. Beyond the scientific relevance, the uncertainty of gas, water, electricity, fuel meters, scales, etc. are strictly regulated in all countries to protect the customers [19, 20]. The specific heat capacity can be measured within a few percent uncertainty for solid materials [21, 22] and complex liquids like nanofluids [23, 24] as well. Since thermal properties, like thermal conductivity, are often required in chemical engineering, it is usually published in a parameter range which exceeds typical engineering use [25]. The available measurement techniques for thermal conductivity is reviewed by Palacio et al. [26]. Overall, this property can usually be measured also within a few percent uncertainty [27, 28]. Even density measurements are burdened with a similar few percent uncertainty, however, careful techniques could give a result well below 1% [29]. As a general conclusion from the references above, if the material is given, its thermal properties can be measured within a few percent uncertainty. However, when a building has to be designed for various weather conditions [30], the overall uncertainty will be extremely high due to the weather conditions to be considered. Nevertheless, the problem has to be solved.

3.1 Calibration

Selecting the appropriate measurement device for an application in thermal engineering is always a struggle when it is not a standard product like thermometers or pressure transmitters. The price range is usually high which correlates with the measurement range and accuracy of the device. Above a certain market value, a calibration sheet also comes with the device which means that it is tested to a standard quantity in few or several points. This is a form of quality control which is advised for all measurements. Obviously, the effort required for calibration is directly charged to the final price of the device. Note that calibration should be repeated regularly. Depending on the environment and device use, the interval can spread from multiple times a day to once per year. Therefore, it is often impractical to ask the vendor every time to do it for you.

If a device comes without factory calibration, information on accuracy is usually given. It can be a few percents of its full-scale limitation, its reading, or a given quantity. Even though these values do not seem high at first glance, if the quantity to be calculated relies on three-four of these measurement results, the final accuracy will easily exceed 10–15% that is often excessive. Hence, the demand for calibration rises. On the other hand, when a device arrives with a factory calibration sheet, it does not necessarily match the operating range of the application, or simply the environment will be different. As an example, a flow meter which was calibrated with water needs recalibration when used with rapeseed oil due to the largely different fluid properties. It is improbable that the manufacturer can be asked to use an alternative liquid for this procedure. Hence, a knowledge on calibration is always useful to have reliable measurement data. A practice-oriented review on the calibration of various industrial sensors can be found in Ref. [31]. The procedure for various cases with practical advices was summarized by Taylor and Oppermann [32]. See Ref. [33] for the Monte Carlo Method, Ref. [34] for nonlinear sensors, and Ref. [35] for an overview on calibration and uncertainty. In the followings, the calibration procedure is presented for a K-type thermocouple, assuming a normal distribution of the samples, hence, the Student's t-distribution can be used.

The calibration procedure starts with setting up a calibrator which will provide the well-controlled quantities to be measured by a sensor. Then several individual points are set and the sensor data are recorded. Depending on the sensor type, ramping up and down might be useful to account hysteresis. The selection of measurement points can be made linearly for linear sensors. Note that an excessive calibration range will lead to higher uncertainty, therefore, it is advised to calibrate only for the range of use. Upon completing the measurement, the evaluation procedure starts. The first step in the present case is fitting a line to the data which approximates the real behavior:

$$U = a \cdot T + b \approx U_{fit} = \alpha \cdot T + \beta, \tag{1}$$

where U is the measured electric voltage caused by the Seebeck effect, T is the temperature, and parameters a and b are approximated by α and β. These are calculated by Eqs. 2 and 3, respectively:

$$\alpha = \frac{\sum_j U_j \cdot (T_j - \bar{T})}{\sum_j (T_j - \bar{T})}, \tag{2}$$

$$\beta = \bar{U} - \alpha \cdot \bar{T}, \tag{3}$$

where overbar means the mean value.

Then the confidence interval is calculated. Firstly, the residual standard deviation is determined for $n - 2$ degrees of freedom, where n is the total number of samples:

$$\sigma_{res}^2 = \frac{1}{n-2} \sum_j \left(U_j - \alpha \cdot T_j - \beta\right)^2. \tag{4}$$

The standard deviation of U_{fit} at the jth point:

$$\sigma_{fit,j}^2 = \frac{\sigma_{res}^2}{n} + \sigma_{res}^2 \cdot \frac{(T_j - \bar{T})^2}{\sum_j (T_j - \bar{T})^2}. \tag{5}$$

Lastly, the inverse of the Student's t-distribution, λ_{st}, is determined, using a lookup table. Its input parameters are the probability and the degree of freedom, which is $n - 1$ here. Hence, the confidence interval can be calculated as:

$$\gamma_j = \lambda_{st} \cdot \sigma_{fit,j}, \tag{6}$$

substituting γ_j into the fitted equation:

$$U_{conf,j} = \alpha \cdot T_j + \beta \pm \gamma_j. \tag{7}$$

Figure 3 shows the result of the calibration procedure. Confidence intervals were calculated for 95% probability, and their values were multiplied by 50 for better visibility. Without this, the three curves would overlap each other which is otherwise desired. During measurements, the inverse of the calibration diagram is used since the temperature has to be calculated by using the measured voltage.

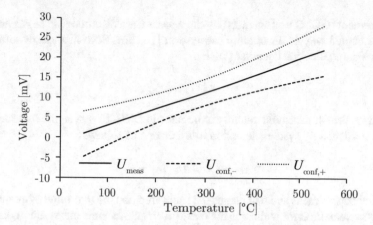

Fig. 3 Results of calibration with confidence intervals multiplied by 50

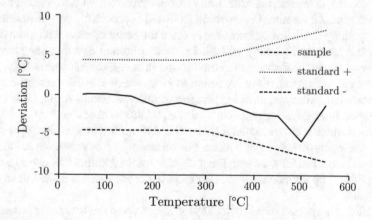

Fig. 4 Comparison of the result of calibration with the thermocouple uncertainty standard

Since K-type thermocouples have a standard accuracy of $\max(2.2\,^\circ\mathrm{C},\,0.75\%)$ at $1 \cdot \sigma$—note that the latter limitation is understood as 0.75% of the temperature in $^\circ\mathrm{C}$, not in K—, the calibration results can be compared to this limitation. The result is shown in Fig. 4 at $2 \cdot \sigma$. Note that the deviation from the standard values originated from all uncertainties present in the measurement loop, including the data acquisition system, not only from the thermocouple itself.

3.2 Uncertainty Estimation

The classical way of uncertainty estimation is the use of the Taylor Series Method (TSM). However, in 2008, the Guideline for the expression of Uncertainty in

Measurement (GUM) was amended by the Monte Carlo Method (MCM) as an additional accepted way of uncertainty estimation [18, 36]. Both methods assume that there is a data reduction equation (DRE):

$$r = r\,(x_1, ..., x_n)\,,\tag{8}$$

where x_i is the ith measured quantity to be used in DRE. However, the measurement of x_i is burdened by systematic and random errors, as follows:

$$x_{i,meas} = x_{i,real} + \beta_{x_i} + \varepsilon_{x_i},\tag{9}$$

where subscripts meas and real means the measured and the real value of quantity x_i. β is the systematic error while ε is the random error. Since the errors are understood as random variables, $x_{i,meas}$ will also be a random variable. According to the Central Limit Theorem, if several, not necessarily normal Probability Distribution Functions (PDF) are summed together, the resulting PDF will approach the normal distribution. Since a quantity that a sensor measures is often the result of several phenomena, the PDF of the measured quantity will likely show a normal distribution. This is the reason why normal distribution is emphasized in measurement technology. When uncertainty is mentioned, it usually refers to $1 \cdot \sigma$, where σ is the standard deviation of the normal distribution, meaning that the quantity is certain within 68% level of significance. In engineering practice, $2 \cdot \sigma$, i.e., 95.4% level of significance, is generally desired which is also recommended by GUM [36]. However, critical applications and precise sensing might require more. For instance, $5 \cdot \sigma$ is a standard value in particle physics to consider a measurement significant. In addition, the experiment has to be validated by an independent institution to accept and use the result in model development.

In TSM, r is approximated by its Taylor Series up to the first order around r_{real}. The measurement uncertainty is understood in practice as the standard deviation of $\delta_r = r_{meas} - r_{real}$ multiplied a number, according to the desired level of significance.

$$\sigma_{\delta_r} = \sqrt{\sum_{i,j}\left[\frac{\partial r}{\partial x_i} \cdot \frac{\partial r}{\partial x_j} \cdot \sigma_{x_i} \cdot \sigma_{x_j} \cdot \mathrm{corr}\,(x_i, x_j)\right]},\tag{10}$$

where corr $\in [-1, 1]$ is the correlation between two quantities. If the measurement of all quantities is independent of each other, Eq. 10 simplifies into the most widely used expression in engineering for uncertainty estimation, Eq. 11:

$$\sigma_{\delta_r} = \sqrt{\sum_i\left(\frac{\partial r}{\partial x_i} \cdot \sigma_{x_i}\right)^2}.\tag{11}$$

Conclusions from Eqs. 10 and 11:

- If all the measured quantities are independent of each other, i.e., corr $(x_i, x_j) = 0$ when $i \neq j$, the resulting uncertainty will always increase if more quantities are considered in r.
- Neglecting corr (x_i, x_j) is not a safe approximation. Since both this quantity and the multiplication of the mixed partial derivatives can be either positive or negative, these terms may lead to either increased or decreased uncertainty.
- Awareness of the corr (x_i, x_j) term can be used to lower the measurement uncertainty. For instance, when the temperature difference is measured, and the two thermometers were calibrated against the same reference thermometer together, the uncertainty of the measurement can be reduced into a fraction of the uncertainty of either thermometer, including the reference thermometer as well. See Example 6-2.4 in Ref. [36] for details.

In MCM, the simplifications made in TSM can be omitted. The DRE is directly used, putting Eq. 9 into Eq. 8. Here, all systematic and random errors as random variables needs to be fully known. The algorithm is the following.

1. Put all guessed $x_{i,real}$ and all random variables into the DRE.
2. Put a randomly chosen variable from systematic and random error PDFs into the DRE M times which should be high enough that usually means thousands or even more.
3. The calculated DRE values form a random variable which can be directly used for uncertainty estimation.

The above procedure is straightforward for uncorrelated variables. If corr (x_i, x_j) $\neq 0$ when $i \neq j$, and all the PDFs are normal, Cholesky transformation of the correlation matrix will reveal the parameter dependence, and the calculation will become matrix algebra. For non-normal distributions, copulas [37] are recommended to use for randomly drawing the samples since $\sigma_{x_i} \cdot \sigma_{x_j} \cdot$ corr (x_i, x_j) will not work. Practical details with examples can be found in Ref. [38].

3.3 What to Do Without Proper Material or Measurement Data or Known Thermal Conditions?

In the classroom, thermal problems are always introduced with given material properties with zero uncertainty, and the case to be solved contains isotropic materials. For steady-state analysis, only the thermal conductivity of the material is the necessary property, which is a single value. In transient cases, specific heat capacity and density are extra parameters. In reality, if the metal is not structural steel or aluminum, and the fluid is not water or air, the struggle comes how to get proper data that is quite usual in thermal engineering. With more guessing, the result of the model will be less certain, if usable in any extent. Note that polymers often show a different behavior, even when the material type is identical, e.g., due to one is in amorphous form while the other has a crystal structure or a combination of the two forms. Material properties are typical parameters that an engineer learns through his/her career and will be able to give excellent guesses after several years of experience.

This is the point when the thermal engineer must consult the situation with the leaders and find the best strategy to solve the issue. The available solutions are the following in order, summing the required effort and solution cost together.

1. Searching for the missing data in free online databases, like National Institute of Standards and Technology [39], Dortmund Data Bank [40], and Engineering ToolBox [41]. This last source is also rich in illustrations, features calculators, and several practical tools to support everyday engineering work.
2. Purchasing entry to online databases. This requires attention since it is not guaranteed that you really get the material data what you are looking for. In addition, several databases exist which actually only mirror data from free databases while asking money for access. Therefore, a guarantee is necessary in advance to avoid inconvenient situations.
3. If the material of interest is missing from a database, properties of a similar material can be used for estimations. However, this decision must be revisited later on.
4. When properties of liquids or gases are missing, it is sure that there are estimating methods developed several decades ago by principally chemical engineers. Reference [42] provides an overview of these methods. Be careful with these estimations since they might bear an undesired deviation [43].
5. Ask a thermal lab to perform the necessary measurements for you. At first, it seems a waste of money, however, they are experts in their field, and they surely consider all notable conditions that might bias the measurement results of a homemade experiment.
6. Perform measurements on your own by using samples. Since thermometers are among the cheapest sensors, it might feel easy to do the measurements by yourself. Then the realization will come that avoiding all biasing effects is rather hard. Even more, without calibration, you can only believe that the purchased sensors show the right value.
7. If the final product or a prototype exists, perform measurements on that and try to determine the missing material properties by reverse engineering. This requires a solid theoretical background and experience in thermal engineering since, e.g., an omitted thermal radiation promptly leads you to false results.

Before using any data, know its uncertainty, discussed in Sect. 3. Unfortunately, this information is rarely discussed in the literature. However, if the measurement technique used for the measurement is known, then the uncertainty can be roughly estimated [44]. In addition to the above tips, it is recommended for all thermal engineers to have the following sensors ready to take with you to any site. Manual reading variants are good, however, an increasing number of measurement devices feature smartphone connectivity, hence, offering online data logging as well.

- A touch probe K-type thermocouple which can be used in the range of -270–$1260\,°C$ temperature with appropriate handle design.

- A thermal camera for qualitative evaluation. Note that various materials might feature various emissivities which may lead to highly biased results, hence, do not make decisions only on thermal images. The touch probe will help you in determining the appropriate emissivity value, and then the thermal image will become a quantitative result for an object/feature with uniform surface finish.
- Pitot-static tube with a differential pressure sensor for velocity and pressure measurement. For low flow rates, an anemometer is a better option.

References

1. H. Struchtrup, *Thermodynamics and Energy Conversion* (Springer, Heidelberg, 2014)
2. M. Martin, K. Holge (eds.), *VDI Heat Atlas*, 2nd edn. (Springer, Berlin, 2010)
3. M. Pękal, M. Wojtyra, J. Frączek, Free-body-diagram method for the uniqueness analysis of reactions and driving forces in redundantly constrained multibody systems with nonholonomic constraints. Mech. Mach. Theory **133**, 329–346 (2019)
4. J. Kövecses, Dynamics of mechanical systems and the generalized free-body diagram-Part I: general formulation. J. Appl. Mech. **75**(6), 061012 (2008)
5. J.P. Barreto, A. Trigo, P. Menezes, J. Dias, A.T. De Almeida. FED-the free body diagram method. Kinematic and dynamic modeling of a six leg robot, in *Proceedings of the 5th International Workshop on Advanced Motion Control* (1998), pp. 423–428
6. H. Bao, N.T. Dinh, J.W. Lane, R.W. Youngblood, A data-driven framework for error estimation and mesh-model optimization in system-level thermal-hydraulic simulation. Nucl. Eng. Des. **349**(February), 27–45 (2019)
7. M. Cavazzuti, G. Gaspari, S. Pasquale, E. Stalio, Thermal management of a formula E electric motor: analysis and optimization. Appl. Therm. Eng. **157**(March), 113733 (2019)
8. M. Leśko, W. Bujalski, K. Futyma, Operational optimization in district heating systems with the use of thermal energy storage. Energy **165**, 902–915 (2018)
9. M. Poole, R. Murray, S.M. Davidson, P.D. Docherty, The quadratic dimensional reduction method for parameter identification. Commun. Nonlinear Sci. Numer. Simul. **73**, 425–436 (2019)
10. Y. Wang, J. Huang, C. Su, H. Li, Furnace thermal efficiency modeling using an improved convolution neural network based on parameter-adaptive mnemonic enhancement optimization. Appl. Therm. Eng. **149**(August 2018), 332–343 (2019)
11. F. Ghani, R. Waser, T.S. O'Donovan, P. Schuetz, M. Zaglio, J. Wortischek, Non-linear system identification of a latent heat thermal energy storage system. Appl. Therm. Eng. **134**(January), 585–593 (2018)
12. X. Li, S. Lin, J. Zhang, T. Zhao, Model parameter identification of indoor temperature lag characteristic based on hysteresis relay feedback control in VAV systems. J. Build. Eng. **25**(March), 100839 (2019)
13. M. Baranski, J. Fütterer, D. Müller, Distributed exergy-based simulation-assisted control of HVAC supply chains. Energy Build. **175**, 131–140 (2018)
14. M.S. Murshitha Shajahan, D. Najumnissa Jamal, V. Aparna, Controller design using quantitative feedback theory for thermal power plant process. Case Stud. Therm. Eng. **14**(April), 100441 (2019)
15. A. López, W. Nicholls, M.T. Stickland, W.M. Dempster, CFD study of Jet Impingement Test erosion using Ansys Fluent® and OpenFOAM®. Comput. Phys. Commun. **197**, 88–95 (2015)
16. G. Lomonaco, W. Borreani, M. Bruzzone, D. Chersola, G. Firpo, M. Osipenko, M. Palmero, F. Panza, M. Ripani, P. Saracco, C. Maria Viberti, Initial thermal-hydraulic assessment by OpenFOAM and FLUENT of a subcritical irradiation facility. Therm. Sci. Eng. Prog. **6**(October 2017), 447–456 (2018)

17. P. Olmeda, A. Tiseira, V. Dolz, L.M. García-Cuevas, Uncertainties in power computations in a turbocharger test bench. Meas. J. Int. Meas. Confed. **59**, 363–371 (2015)
18. H. Carstens, X. Xia, S. Yadavalli, Measurement uncertainty in energy monitoring: Present state of the art. Renew. Sustain. Energy Rev. **82**(August 2017), 2791–2805 (2018)
19. European Parliament. Directive 2004/22/EC on measuring instruments (2004)
20. J. Lansing, AGA-9 measurement of gas by multipath ultrasonic meters. Technical Report, SICK Maihak, Inc. (2007)
21. D.W. Waples, J.S. Waples, A review and evaluation of specific heat capacities of rocks, minerals, and subsurface fluids. Part 2: Fluids and porous rocks. Nat. Resour. Res. **13**(2), 123–130 (2004)
22. D. Basak, R.A. Overfelt, D. Wang, Measurement of specific heat capacity and electrical resistivity of industrial alloys using pulse heating techniques. Int. J. Thermophys. **24**(6), 1721–1733 (2003)
23. B. Barbés, R. Páramo, E. Blanco, C. Casanova, Thermal conductivity and specific heat capacity measurements of CuO nanofluids. J. Therm. Anal. Calorim. **115**(2), 1883–1891 (2014)
24. S.M. Sohel Murshed, Simultaneous measurement of thermal conductivity, thermal diffusivity, and specific heat of nanofluids. Heat Transf. Eng. **33**(8), 722–731 (2012)
25. B. Le Neindre, P. Desmarest, G. Lombardi, M. Kayser, G. Chalvignac, F. Gumerov, Y. Garrabos, Thermal conductivity of gaseous and liquid n-pentane. Fluid Phase Equilibria **460**, 146–154 (2018)
26. A. Palacios, L. Cong, M.E. Navarro, Y. Ding, C. Barreneche, Thermal conductivity measurement techniques for characterizing thermal energy storage materials - A review. Renew. Sustain. Energy Rev. **108**(March), 32–52 (2019)
27. R.M. Sarviya, V. Fuskele, Review on Thermal Conductivity of Nanofluids. Mater. Today Proc. **4**(2), 4022–4031 (2017)
28. A. Lorenzati, S. Fantucci, A. Capozzoli, M. Perino, VIPs thermal conductivity measurement: test methods, limits and uncertainty. Energy Procedia **78**, 418–423 (2015)
29. A. Pustogvar, A. Kulyakhtin, Sea ice density measurements. Methods and uncertainties. Cold Reg. Sci. Technol. **131**, 46–52 (2016)
30. W. Tian, Y. Heo, P. de Wilde, Z. Li, D. Yan, C. Soo Park, X. Feng, G. Augenbroe, A review of uncertainty analysis in building energy assessment. Renew. Sustain. Energy Rev. **93**(January 2017), 285–301 (2018)
31. M. Cable, *Calibration: A Technician's Guide* (International Society of Automation, Durham, 2005)
32. J.K. Taylor, H.V. Oppermann, *Handbook for the Quality Assurance of Metrological Measurements.* (U.S. Department of Commerce, National Bureau of Standards, Gaithersburg, 1986)
33. P.L. Bonate, Approximate confidence intervals in calibration using the bootstrap. Anal. Chem. **65**(10), 1367–1372 (1993)
34. L.M. Schwartz, Nonlinear calibration. Anal. Chem. **49**(13), 2062–2069 (1977)
35. C.F. Dietrich, *Uncertainty, Calibration and Probability*, 2nd edn. (Taylor & Francis, Boca Raton, 1991)
36. H.W. Coleman, W.G. Steele, *Experimentation, Validation, and Uncertainty Analysis for Engineers*, 4th edn. (Wiley, New York, 2018)
37. H. Joe, *Dependence Modeling with Copulas* (Chapman and Hall/CRC, Boca Raton, 2015)
38. Mathworks. Simulating Dependent Random Variables Using Copulas (2019)
39. National Institute of Standards and Technology, Material Measurement Laboratory (2019). https://www.nist.gov/mml
40. Dortmund Data Bank (DDB). www.ddbst.com
41. Engineering ToolBox, https://www.engineeringtoolbox.com/ (2019)
42. B.E. Poling, J.M. Prausnitz, J.P. O'Connell, *The Properties of Gases and Liquids*, 5th edn. (McGraw-Hill, New York, 2001)
43. D. Csemány, V. Józsa, Fuel evaporation in an atmospheric premixed burner: sensitivity analysis and spray vaporization. Processes **5**(4), 80 (2017)
44. C. Olm, T. Varga, É. Valkó, S. Hartl, C. Hasse, T. Turányi, Development of an ethanol combustion mechanism based on a hierarchical optimization approach. Int. J. Chem. Kinet. **48**(8), 423–441 (2016)

Chapter 2
General Aspects of Thermodynamical Modeling

1 Onset of Depression

Thermal processes are natural phenomena, not restricted to engineering applications. It is essential to discuss the way how the mathematical models are built, what is their structure and what part can be exchanged to another in order to fit the description to the given task to solve. This chapter is intended to briefly summarize the most critical aspects of thermodynamical modeling, for details, we refer to the following literature [1–7].

First of all, we must emphasize that the mathematical models are only approximations, based on our imperfect knowledge of reality. It does not matter how much experimental data we have; it remains imperfect. However, one can choose the best-known tool to catch the essence of a problem. The differences between these tools are the following:

1. number of variables,
2. applied constitutive equations,
3. resulting number and type of the equations,
4. solution method,
5. predictive power or accuracy,
6. interpretation and physics behind.

There are connections between these attributes, i.e., they are not independent of each other. For instance, it is not possible to choose a large number of variables with keeping low the number of equations or choose thermodynamically incompatible constitutive laws to describe a coupled phenomenon.

Let us start with the last one. The interpretation of the results and their meaning are entirely determined by the approach used to derive the models. For example, in the kinetic theory, there are a particular set of particles with prescribed properties, interaction models, and this approach highly exploits this specific description of the system. As a result, for instance, the thermal conductivity can be determined based on these prescribed features of the particles and interpreted as a parameter that

© Springer Nature Switzerland AG 2020
V. Józsa and R. Kovács, *Solving Problems in Thermal Engineering*, Power Systems,
https://doi.org/10.1007/978-3-030-33475-8_2

characterizes the interaction process among the particles. On the other hand, it is still possible to choose another interaction model that leads to a different formula for thermal conductivity. Even more important, this approach constrains the possibilities in parallel, that is, a constraint means fixed parameters in the resulting equations, i.e., they cannot be tuned for experimental data. It turns out immediately whether the approach is appropriate or not. Furthermore, there are still other possibilities beyond the kinetic theory, such as the approach of Classical Irreversible Thermodynamics (CIT) [2, 5, 6, 8, 9], that offers a kind of recipe in order to derive new constitutive equations based on the second law of thermodynamics, as a continuum model. However, it gives no clue how to calculate the thermal conductivity, for instance. To do so, one has to perform experiments, and this material parameter can be fitted to particular data. Since the background of these approaches is different, the resulting models are also different, even with the same structure of equations.

The predictive power of an approach can be or should be tested on experiments. This aspect means how precise the model is. Some of them are easy to solve or more straightforward to implement but offer only qualitative predictions instead of quantitative ones. Naturally, the goal is to be as accurate as possible while keeping the number of variables as low as possible. It is also a consequence of the approach, as a result of decisions in the early phase of the modeling.

The mathematical models are mostly ordinary or partial differential equations (ODE or PDE). Their properties, as well as the related boundary and initial conditions, depending on the approach. The kinetic theory sometimes leads to hyperbolic type PDEs with a well-defined structure of equations; however, its size can be enormously large in some instances.[1] On the other hand, the outcome of 'CIT-like' methods depends on the variables; it can be both hyperbolic or parabolic. There are numerous ways how to solve these models; Chap. 6 is devoted to discussing it.

Here we came to the constitutive equations, i.e., to the equations that describe how the material behaves. Their use in the modeling process is inevitable and have a crucial impact on the results. In many situations, the constitutive equations are not separately discussed from the balance equations; however, they are substituted into, and in such a way, their true nature remains hidden. This is problematical when one has to apply a different constitutive equation, without knowing the specific structure behind.

The chosen approach constrains the way how to derive constitutive equations, and also their structure. Moreover, it must respect certain physical principles such as objectivity [11] and the second law of thermodynamics [12] in order to obtain a stable, thermodynamically consistent and physically sound model. Otherwise, the outcome becomes questionable.

Finally, we came to the first step: the applied set of variables. The modeling problem itself restricts some of them; for instance, a thermal problem requires to use the temperature as a primary field variable. However, for a multicomponent system, it is not necessary to use a different temperature for each component, depending on

---

[1] We note here that there are so-called 'regularization techniques' which stabilize the behavior and yield parabolic models [10].

the actual conditions. Indeed, it is a matter of choice. So to say, there is a lower limit for the set of variables that should reflect enough information about the problem, keeping the treatment as simple as possible.

In the following, we are going to overview how the approach of CIT works with a given set of variables, and how it is embedded into a universal generalization method. The first and the second laws of thermodynamics are exploited intensely, and in order to gain a clear picture, this is what we shall discuss now.

The first law of thermodynamics represents the time evolution of internal energy using a balance equation:

$$\rho \partial_t e + \nabla \cdot \mathbf{J}_e = q_v, \tag{1}$$

where e is the specific internal energy and the caloric equation of state ($e(T, \rho)$) connects it to the temperature, and ρ denotes the mass density. The partial time derivative is denoted by ∂_t and ∇ being the well-known nabla operator. The current density of e is $\mathbf{J}_e$ that is reducing to the heat flux in rigid heat conductors, i.e., when the flow of the material does not carry internal energy. Generally speaking, the divergence of $\mathbf{J}_e$ represents the inflow and the outflow from the control volume. On the right-hand side, the term q_v stands for the volumetric production which could be negative, too. The particular form of q_v depends on the nature of the source, such as

- internal heat generation or drain induced by radiation,
- heat convection effects,
- heat generation or drain due to chemical reactions,
- mechanical work (e.g., compressibility),
- electric dissipation (e.g., Joule heating),

can be included, and q_v may depend on time and space as well. For instance, one interesting and challenging problem is to determine the heat generation field in a fusion reactor that essentially influences the operation of the complete system [13–16]. This extraordinary problem very well represents the difficulties arising in the determination of source terms in the balance equations.

For classical continua, the physical content of the second law is the asymptotic stability of homogeneous equilibrium solutions of the field equations [17]. The well-known particular requirements for a suitable mathematical model are the following.

1. The entropy is a concave function of the corresponding state variables.
2. The entropy is increasing in an insulated system, that is the entropy production is non-negative,

therefore the balance of entropy is an inequality:

$$\rho \partial_t s + \nabla \cdot \mathbf{J}_s = \sigma_s \geq 0, \tag{2}$$

where s denotes the specific entropy, $\mathbf{J}_s$ is the entropy current density. In non-dissipative case the entropy production σ_s is zero. The solution to this inequality yields the constitutive equations. Its mathematical formulation is clear since it is free from ambiguities that arise in the classical literature of temporal[2] thermodynamics which is deeply criticized by Matolcsi [1].

The previous balances, (1) and (2), are eligible to model heat conduction phenomenon without any coupling, for instance, to the mechanical effects. However, the approach of CIT is not merely about heat conduction: it is a useful tool to derive constitutive equations, also beyond thermal phenomenon.

In the following, we briefly overview the CIT approach for various situations, in which the local equilibrium hypothesis plays a fundamental role. It assumes that the same state variables characterize the process out of equilibrium as in equilibrium. In contrary, non-equilibrium thermodynamics extends the state space with 'dynamic variables' that vanish in equilibrium. CIT exploits the second law rigorously and considers the balance equations as constraints. Besides the balance of internal energy, one may need to include the mass and momentum balances as well:

$$D_t \rho + \rho \nabla \cdot \mathbf{v} = 0, \tag{3}$$

$$\rho D_t \mathbf{v} + \nabla \cdot \mathbf{P} = \rho \mathbf{f}, \tag{4}$$

in which D_t denotes the substantial time derivative that reduces to the partial time derivative for rigid materials. Equation (4) is also known formally as Cauchy's equation of motion with $\mathbf{f}$ being a body force, $\mathbf{v}$ is the velocity field, and $\mathbf{P} = \mathbf{P}^T$ is the symmetric pressure tensor.[3] The source of mass balance becomes important for multicomponent materials, especially when chemical reactions occur. Classically, the coupling between different fields can be realized through the source terms in the balance equations. In non-equilibrium thermodynamics, and also in other modern approaches, this possibility is significantly extended by means of the extended state space. For further details, we refer to [2, 5].

1.1 Heat Conduction

When one accepts the specific internal energy e as the only variable, then a local equilibrium hypothesis is accepted in parallel. That is, while the temperature T completely characterizes the equilibrium state of the conducting medium thermally, we accept that it is valid for an out-of-equilibrium state, too. However, 'not so far from equilibrium'.[4] The caloric equation of state effectuates the connection between the temperature and the internal energy.[5] The simplest case is

---

[2]Or using the usual convention: 'ordinary' thermodynamics which is about to describe the time evolution without considering any spatial dependence.

[3]Only when the internal moment of momenta is absent.

[4]Strictly speaking, it is meaningless in a mathematical sense until the definition of a 'distance' is given, measuring the 'interval' between states in which one is related to equilibrium state.

[5]In a more general situation, the mass density ρ is also present, i.e., $e = e(T, \rho)$.

$$e = c_v T,\tag{5}$$

where c_v is the isochoric specific heat. In the following, we deal with rigid heat conductors only. Therefore, other specific heats will not play any role. Moreover, we consider only isotropic materials. One needs to use a balance equation for each variable; in this case, it is the balance of internal energy ($q_v = 0$):

$$\rho \partial_t e + \nabla \cdot \mathbf{q} = 0,\tag{6}$$

where $\mathbf{q}$ is called the conductive current density of internal energy, i.e., the heat flux. Here we note that there are other possibilities for the equation of states. It is also chosen for the particular modeling task, and later, some basic examples will be discussed.

The specific entropy depends only on the specific internal energy e, $s = s(e)$. To calculate the entropy production (2), one needs the connection between the entropy and its variables, it is given according to the Gibbs relation [5]:

$$T\,\mathrm{d}s = \mathrm{d}e, \Rightarrow \frac{\mathrm{d}s}{\mathrm{d}e} = \frac{1}{T}.\tag{7}$$

The time derivative of the specific entropy becomes

$$\partial_t s = \frac{\mathrm{d}s}{\mathrm{d}e}\partial_t e = \frac{1}{T}\partial_t e = -\frac{1}{T}\nabla \cdot \mathbf{q},\tag{8}$$

using the chain rule according to Eq. (7), and the balance equation, too. The usual entropy current density $\mathbf{J}_s$ is

$$\mathbf{J}_s = \frac{1}{T}\mathbf{q}.\tag{9}$$

It is not the only possibility for $\mathbf{J}_s$, there exist other choices in the literature, depending on the level of generalization, showing examples in the Sect. 2. Then, the entropy production is

$$\sigma_s = -\frac{1}{T}\nabla \cdot \mathbf{q} + \frac{1}{T}\nabla \cdot \mathbf{q} + \mathbf{q} \cdot \nabla \frac{1}{T} = \mathbf{q} \cdot \nabla \frac{1}{T} \geq 0.\tag{10}$$

Following the work of Onsager [18, 19], the inequality of entropy production has the general structure

$$\sigma_s = \sum_i X_i J_i \geq 0,\tag{11}$$

where X_i are called thermodynamical forces and J_i are the thermodynamical fluxes. In this case, the thermodynamical flux is the heat flux and the thermodynamical force is not the temperature gradient but the gradient of the reciprocal temperature. It is especially important for low-temperature problems in order to prevent to reach the 0 K state or going below.

The simplest solution of this inequality assumes a linear relationship between the force and flux such as

$$\mathbf{q} = l\nabla\frac{1}{T}, \tag{12}$$

with $l > 0$ being a constant for simplicity. Again, it is arbitrary to consider this choice of solution. The solution of (11) is the constitutive equation itself, in case of heat conduction, (12) is called Fourier's law, and l is called a thermodynamical conduction coefficient which is not the same as thermal conductivity. The thermal conductivity λ is obtained after performing the differentiation, that is

$$\lambda = \frac{l}{T^2}, \tag{13}$$

so the Fourier's law becomes

$$\mathbf{q} = -\lambda\nabla T. \tag{14}$$

Note that l also could be a temperature dependent coefficient. In general, the T-dependence of the thermal conductivity λ can be measured, and using Eq. (13), the $l(T)$ can be recovered.[6] In summary, a thermal problem without considering any other phenomenon can be solved using the system of balance and constitutive equations:

$$\rho\partial_t e + \nabla \cdot \mathbf{q} = q_v, \tag{15}$$
$$\mathbf{q} = -\lambda\nabla T, \tag{16}$$
$$e = c_v T. \tag{17}$$

Any other form of Eq. (16) is only a consequence of this previous derivation. For instance, in many cases the equation

$$\frac{\partial T}{\partial t} = \alpha\Delta T \tag{18}$$

is referred to be the Fourier heat equation where Δ denotes the Laplacian operator and $\alpha = \lambda/(\rho c_v)$ being the thermal diffusivity. However, it also implies that

1. all the material parameters are constant,
2. there is no heat generation,

---

[6]From a practical point of view, neglecting the T-dependence of λ depends on the material, and theoretically, Eq. (13) is a better model for low temperature situations.

3. the Fourier constitutive equation is valid,
4. the temperature is our primary variable,

so the caloric equation of state remains hidden in the background together with other implications. More importantly, it is only a consequence of several decisions made at different stages of derivation. It also constrains the meaning of boundary conditions. As the temperature T is the primary variable, the following boundary conditions can be defined:

1. Dirichlet (or first-type) boundary condition: the primary variable is given directly, i.e., $T(t)$ is prescribed on the boundary.
2. Neumann (or second-type) boundary condition: the fluxes are prescribed in time, which coincides with the gradient of the primary variable here.
3. Robin (or third-type) boundary condition: a linear combination between the previous two is prescribed. As a particular case, the heat convection on the boundary is given as $\mathbf{q} \cdot \mathbf{n} = -\lambda \nabla T \cdot \mathbf{n} = h(T - T_\infty)$ with T_∞ being the constant temperature of the fluid flow, h is the heat transfer coefficient and $\mathbf{n}$ represents the surface normal vector.

It is not necessary to use the temperature as a primary variable; it is also possible to use

$$\frac{\partial \mathbf{q}}{\partial t} = \alpha \Delta \mathbf{q} \tag{19}$$

as a choice. When the Fourier's law (12) is valid, then there is no any importance of this choice, but it is important for generalized heat conduction laws that discussed later.

1.2 Heat Conduction in Isotropic Fluids: Navier–Stokes–Fourier Equations

Here, we show an example for modeling a heat conducting isotropic fluid. The primary field variables are the specific internal energy e and the mass density ρ. The latter is not constant now. We shall follow the same steps and aspects as previously. Let us recall the balance equations for mass and internal energy, using substantial time derivatives,

$$D_t \rho + \rho \nabla \cdot \mathbf{v} = 0, \tag{20}$$

$$\rho D_t e + \nabla \cdot \mathbf{q} = -\mathbf{P} : \nabla \mathbf{v}. \tag{21}$$

Here, $\mathbf{P}$ is the pressure tensor that can be decomposed into static (p) and dynamic (or viscous, $\mathbf{\Pi}$) parts:

$$\mathbf{P} = \mathbf{\Pi} + p\mathbf{I}, \tag{22}$$

where $\mathbf{I}$ is the identity tensor. There is a source term now in the energy balance due to the compressibility property of the fluid, it expresses the thermal dissipation inside the medium, induced by the mechanical work. The derivation of constitutive equations does not require the momentum balance, but it is required to the complete description of the phenomenon:

$$\rho D_t \mathbf{v} + \nabla \cdot \mathbf{P} = \rho \mathbf{f}. \tag{23}$$

The static pressure p is connected to the temperature through the thermal equation of state $p = p(T, \rho)$ that can be, for instance,

$$p = R\rho T, \tag{24}$$

where R is the specific gas constant, assuming ideal gas. Since $s = s(e, \rho)$, the connection between the specific entropy and its variables is expressed by the Gibbs relation:

$$T\mathrm{d}s = \mathrm{d}e - \frac{p}{\rho^2}\mathrm{d}\rho. \tag{25}$$

Using the same entropy current density $\mathbf{J}_s$ as previously, the calculation of entropy production is straightforward:

$$\sigma_s = -\frac{1}{T}\,(\mathbf{P} - p\mathbf{I}) : \nabla \mathbf{v} + \mathbf{q} \cdot \nabla \frac{1}{T} \geq 0. \tag{26}$$

That is, the dynamical pressure $\mathbf{\Pi}$ appears within the inequality and the constitutive relations, for anisotropic materials, are

$$\mathbf{q} = \mathbf{\Lambda}\nabla\frac{1}{T},$$
$$\mathbf{\Pi} = \frac{1}{T}\mathbf{L}_{(4)}(\nabla \mathbf{v}), \tag{27}$$

where $\mathbf{\Lambda}$ and $\mathbf{L}_{(4)}$ are a symmetric, positive definite second and fourth order tensors, respectively. The thermal conductivity $\mathbf{\Lambda}$ is reducing to a constant λ for isotropic case. Regarding the second equation, assuming again the simplest case, the dynamical pressure is taken to be a linear function of the velocity gradient. The dynamical pressure $\mathbf{\Pi}$ can be decomposed into spherical and deviatoric parts in isotropic materials as

$$\mathbf{\Pi}^{\mathrm{sph}} = \frac{1}{3}\mathrm{Tr}\mathbf{\Pi} = \eta_b \nabla \cdot \mathbf{v}, \tag{28}$$
$$\mathbf{\Pi}^{\mathrm{dev}} = \mathbf{\Pi} - \mathbf{\Pi}^{\mathrm{sph}}\mathbf{I} = \eta_s (\nabla \mathbf{v})_{\mathrm{dev}}, \tag{29}$$

which implies that the pressure tensor $\mathbf{P}$ is symmetric; furthermore, η_s denotes the shear viscosity and η_b is the bulk viscosity [5]. The Newton's law for fluids is obtained as a natural consequence of the second law. The spherical part becomes significant for compressible fluids, for instance, for rarefied gases. We add some further remarks.

1. In the classical literature of fluid mechanics, the momentum equation (4) is often referred to be the Navier–Stokes equation with $\mathbf{\Pi}$ being substituted into. Again, it could be confusing because several assumptions are made to derive the present form of constitutive equations. For non-Newtonian fluids, such substitution would make the modeling task more difficult to solve.
2. There is no direct coupling (cross-effect) between the constitutive equations in isotropic materials. This is due to the representation theorems of isotropic functions [4], commonly called Curie principle. In classical case, the coupling can be realized only through the balance equations.
3. The balance of momentum was an indirect constraint through the internal energy [20].

In summary, the complete system of Navier–Stokes–Fourier equations reads

$$\mathrm{D}_t\rho + \rho\nabla \cdot \mathbf{v} = 0,$$
$$\rho\mathrm{D}_t e + \nabla \cdot \mathbf{q} = -\mathbf{P} : \nabla\mathbf{v},$$
$$\rho\mathrm{D}_t\mathbf{v} + \nabla \cdot \mathbf{P} = \rho\mathbf{f}.$$
$$\mathbf{q} = -\lambda\nabla T,$$
$$\mathbf{\Pi}^{\mathrm{sph}} = \eta_b\nabla \cdot \mathbf{v}, \tag{30}$$
$$\mathbf{\Pi}^{\mathrm{dev}} = \eta_s(\nabla\mathbf{v})_{\mathrm{dev}}, \tag{31}$$

together with the caloric ($e(T, \rho)$) and thermal ($p(T, \rho)$) equations of state. This representation emphasizes the structure of the model and eases the numerical solutions by separating the balances from the constitutive laws. It is especially important for non-classical (generalized) models.

1.3 Mechanics: Examples for Caloric Equation of State

In this subsection, we restrict ourselves to one-dimensional, small elastic deformations in order to avoid problems with objectivity and plasticity. The elastic strain ε is related to the Cauchy (elastic) stress σ_{el} as

$$\sigma_{\mathrm{el}} = E\varepsilon \tag{32}$$

with $E > 0$ being the constant Young modulus. The ideal elasticity is a nondissipative process thus it does not contribute to the entropy production. However, it has a contribution to the specific internal energy, without taking the thermal expansion into account [21–24]:

$$e(T, \varepsilon) = e_{\text{th}}(T) + e_{\text{el}}(\varepsilon),\tag{33}$$

where the thermal part $e_{\text{th}}(T)$ could be the same function as previously and the elastic part $e_{\text{el}}(\varepsilon)$ is

$$e_{\text{el}}(\varepsilon) = \frac{E}{2\rho}\varepsilon^2.\tag{34}$$

This contribution also appears in the balance of internal energy. Our basic field variables are e and ε, i.e., $s = s(e, \varepsilon)$ and the corresponding Gibbs relation is

$$T\mathrm{d}s = \mathrm{d}e - \frac{\sigma_{\text{el}}}{\rho}\mathrm{d}\varepsilon.\tag{35}$$

Repeating the calculation of entropy production, it becomes apparent how the elastic part vanishes:

$$\sigma_s = \frac{\rho}{T}\partial_t e - \frac{\sigma_{\text{el}}}{T}\partial_t\varepsilon + \nabla \cdot \mathbf{J}_s \geq 0,\tag{36}$$

$$\partial_t e = \partial_t e_{\text{th}}(T) + \partial_t e_{\text{el}}(\varepsilon) = \partial_t e_{\text{th}}(T) + \frac{\sigma_{\text{el}}}{\rho}\partial_t\varepsilon,\tag{37}$$

as it is, only the heat conduction part remains in Eq. (36) that leads to the Fourier equation again. Here, the quantity $\partial_t\varepsilon$ is related to the velocity gradient in three dimensions. Moreover, for small deformations, it is possible to approximate ρ with a constant. The Navier–Stokes–Fourier equations are recovered, keeping the sign convention $\mathbf{P} = -\sigma$ in our mind.

So far, the thermal expansion was neglected. Let us now include it as well. Hence the specific internal energy becomes [25]

$$e = e_{\text{th}}(T) + e_{\text{el}}(\varepsilon) + \frac{E\alpha}{\rho}T\varepsilon,\tag{38}$$

where the last term indicates the effect of thermal expansion. Here, α being the thermal expansion coefficient. Using this formulation, the thermal expansion has a contribution in the velocity gradient. Furthermore, as a reversible process, it vanishes from the entropy production due to the same reasons as above. For the objective treatment of thermomechanics, we refer to the work of Fülöp [23].

Previously, it was highlighted how important is to handle separately the balance and constitutive equations. It is especially true in this case, omitting the complete derivation that can be found in [25], the linear model for thermal expansion together with Fourier heat conduction using only the temperature as a field variable becomes:

$$\partial_{tt}(\gamma_1\partial_t T - \lambda\Delta T) = v_{el}^2\Delta\left[\left(\gamma_1 + \frac{(E\alpha)^2 T_0}{\rho v_{el}^2}\right)\partial_t T - \lambda\Delta T\right],\tag{39}$$

with T_0 being a reference temperature, v_{el} is the longitudinal elastic wave propagation speed and $\gamma_1 = \rho c_v - 3E\alpha^2 T_0$. It would be difficult to recover the balance and constitutive relations using this equation. Besides, the meaning of boundary conditions is not apparent and is still the subject of recent researches [25]. So to say, the models both mathematically and physically can be equivalent; however, it is not advantageous to eliminate the variables and using only the temperature, especially in coupled problems.

1.4 Multicomponent Systems: Fick's Law for Diffusion

Here we assume the simplest case: the heat conduction and mechanical effects are absent in the n-component non-reactive, isothermal isotropic fluid, only the diffusion phenomenon is taken into account [6]. That is, we shall define the concentration c_k as

$$c_k = \frac{\rho_k}{\rho}, \quad \sum_k c_k = 1, \quad \rho = \sum_k \rho_k, \quad k = 1, \ldots, n, \tag{40}$$

with ρ_k being the mass density of the kth component, moreover, ρ is constant ($d\rho = 0$). The concentrations c_k are the basic field variables, their balance equations read as

$$\rho \partial_t c_k + \nabla \cdot \mathbf{J}_k = 0, \quad k = 1, \ldots, n, \tag{41}$$

where $\mathbf{J}_k$ is the mass flux of the kth component. As it is a non-reactive fluid, there is no source term in Eq. (41). The corresponding Gibbs relation is

$$T ds = - \sum_k \mu_k dc_k, \tag{42}$$

where μ_k is the chemical potential of the kth component, hence we have $s = s(c_k)$ and the entropy current density becomes

$$\mathbf{J}_s = -\frac{1}{T} \sum_k \mu_k \mathbf{J}_k. \tag{43}$$

From now on, the calculation of entropy production is straightforward:

$$\sigma_s = -\frac{1}{T} \sum_k \mu_k \rho \partial_t c_k - \frac{1}{T} \sum_k \mu_k \nabla \cdot \mathbf{J}_k - \frac{1}{T} \sum_k \mathbf{J}_k \cdot \nabla \mu_k =$$

$$= \frac{1}{T} \sum_k \mu_k \nabla \cdot \mathbf{J}_k - \frac{1}{T} \sum_k \mu_k \nabla \cdot \mathbf{J}_k - \frac{1}{T} \sum_k \mathbf{J}_k \cdot \nabla \mu_k =$$

$$= -\frac{1}{T} \sum_k \mathbf{J}_k \cdot \nabla \mu_k \geq 0. \tag{44}$$

The solution of the inequality (44) is the Fick's law and for an n-component fluid it reads as

$$\mathbf{J}_k = -\frac{1}{T}\sum_j L_{kj}\nabla\mu_j, \quad k = 1,\ldots,n, \tag{45}$$

where L_{kj} is the Onsagerian conduction matrix, containing all the conduction coefficients. Now, we must make the following remark. Here all the mass fluxes have the same tensorial order and therefore, the coupling between these fluxes must be considered, as such, each flux is coupled to all the others. The final form of the Fick's law is

$$\mathbf{J}_k = -\rho\sum_j D_{kj}\nabla c_j, \quad k = 1,\ldots,n, \tag{46}$$

where D_{kj} is the diffusion coefficient matrix that defined as [26]

$$\rho D_{kj} = \frac{1}{T}\sum_l L_{kl}\frac{\partial\mu_l}{\partial c_j}, \quad k = 1,\ldots,n. \tag{47}$$

Seemingly, the mathematical structure of Fourier and Fick laws are the same because

$$\partial_t c = D\Delta c, \tag{48}$$

that holds for a 1-component case and D is analogous to the thermal diffusivity. However, it shadows the most important difference between the thermal and mass diffusion: the mass density appears together with the diffusion coefficient in the constitutive equation. It does not hold for the Fourier's law.

1.5   Coupled Heat and Mass Transport

Now we extend the previous set of variables with the specific internal energy: $s = s(e, c_k)$, thus the Gibbs relation is

$$T\mathrm{d}s = \mathrm{d}e - \sum_k \mu_k \mathrm{d}c_k, \tag{49}$$

which is valid also with the previous conditions. One important simplification is to take the same temperature for all components. Otherwise k-different T_k would appear in the Gibbs relation and also in the constitutive equations. Exploiting the previous definitions, the corresponding balance equations of the densities are the following:

$$\rho \partial_t e + \nabla \cdot \mathbf{q} = 0,$$
$$\rho \partial_t c_k + \nabla \cdot \mathbf{J}_k = 0. \tag{50}$$

Since the heat conduction phenomenon is included, the entropy current density is modified:

$$\mathbf{J}_s = \frac{1}{T} \left(\mathbf{q} - \sum_k \mu_k \mathbf{J}_k \right), \tag{51}$$

and omitting the details of further calculation, the entropy production[7] is

$$\sigma_s = \mathbf{q} \cdot \nabla \frac{1}{T} - \sum_k \mathbf{J}_k \cdot \nabla \frac{\mu_k}{T} \geq 0. \tag{52}$$

As a consequence of their same tensorial rank, the heat flow is coupled to the mass flow, i.e.,

$$\mathbf{q} = l \nabla \frac{1}{T} - \sum_k L_{qk} \nabla \frac{\mu_k}{T},$$
$$\mathbf{J}_k = L_{kq} \nabla \frac{1}{T} - \sum_j L_{kj} \nabla \frac{\mu_j}{T}, \tag{53}$$

where $l > 0$ is the usual part of thermal conductivity, and now the mass flux also has a contribution to the heat flux, expressed through the matrix L_{qk}. Moreover, as the temperature is not constant in this case, it also appears under the gradient. The coefficient matrix L_{kq} expresses the contribution of the temperature gradient to the mass flux of each component. These cross couplings bear the name of Soret and Dufour [5, 6], more precisely:

1. Dufour effect: the heat flux occurs due to an existing concentration gradient,
2. Soret effect: the mass flux occurs due to an existing temperature gradient.

Distinguishing the temperature for all components would lead to a complicated model in which there are k-different heat and mass fluxes with couplings between all components,

$$\mathbf{q}_k = \sum_j l_{kj} \nabla \frac{1}{T_j} - \sum_q L_{kq} \nabla \frac{\mu_q}{T_j},$$
$$\mathbf{J}_k = \sum_q L_{kq} \nabla \frac{1}{T_q} - \sum_j L_{kj} \nabla \frac{\mu_j}{T_q}, \tag{54}$$

---

[7]We note here that a different representation is also possible since the $\nabla \frac{1}{T}$ terms can be unified. In this way, the resulting coefficients will be different, too [27].

that is, for an n-component mixture, there are $2n$ number of fluxes with $2n^2$ couplings. Such a model becomes unsolvable for large n and carries a significant uncertainty in the coupling coefficients.

The thermo-electric effect is thermodynamically analogous with the thermo-diffusion theory, the cross-effects are appearing in the same way. Their names are Seebeck and Peltier effects. The last section in Chap. 4 is devoted to discuss that topic from practical point of view.

2   Extended Material Models

It is crucial to emphasize that all the previously derived models have a specific limit that is hard to define precisely when the local equilibrium hypothesis loses its validity, and a non-equilibrium description is needed. Beyond this limit, some extension of the constitutive equation is required. The way of extension (or generalization) depends on the underlying theory, and as before, it has severe consequences on the resulted model. In the following, we will show some examples of generalized models and about their possible applications. About the basics of non-equilibrium thermodynamics, we refer to the following literature [2, 3, 5, 9].

2.1   Extensions of Fourier Equation

A widely known problem with Fourier's law is its parabolic nature from a mathematical point of view. Strictly speaking, it predicts infinite speed for propagation that can be easily illustrated with the solution of an initial value problem, called Green's function of the Fourier heat equation. Let us assume an infinite one-dimensional space with a unit disturbance that may be represented by a Dirac distribution at $x = 0$ and $t = 0$. Then the solution for $t > 0$ is

$$T(x, t) = \frac{1}{2\sqrt{\pi \alpha t}} e^{-\frac{x^2}{4\alpha t}}, \tag{55}$$

where for any $|x|$ and $t > 0$ the $T(x, t) > 0$ follows. It implies immediate temperature change after an initial disturbance at any distance. It seems to be contradictory to our common physical sense. Despite this parabolic property, the Fourier equation is the most widely applied heat conduction model in the engineering practice. For a more detailed discussion, we refer to the work of Fichera [28].

The propagation speed is one of the central problems in heat conduction. The hyperbolic models that describe finite velocities seem to be more acceptable, and many modeling approaches developed in a way to respect that attribute strictly. For instance, the kinetic theory motivated approaches such as the Extended Irreversible Thermodynamics (EIT) [29–32], and Rational Extended Thermodynamics (RET) [7,

33] are constructed in a way to obtain hyperbolic models. Both theories exploit the kinetic theory in the background; however, on a different level. While RET considers its principles more rigorously, EIT loosens the constrains to keep the modeling on a more general level. As a consequence, it is more natural to derive parabolic models based on EIT.

There is one, even more general approach, called Non-Equilibrium Thermodynamics with Internal Variables (NET-IV) [3]. Its universality originates in internal variables and current multipliers [34, 35] since solely their tensorial order is restricted at the beginning. Using an extended state space, it is possible to find compatibility with RET, EIT, and beyond [36–41]. Those models arise as a special case of a particular set of internal variables, for details, see [42, 43].

It is worth mentioning the other approaches such as GENERIC [44–49], and phase-field theories [50–53]. They all have a particular structure that distinguishes them from the others and interprets each phenomenon differently.

The differences point far beyond than simply describing distinct propagation speeds and possessing different terms in the equations. It is a result of the apriori assumptions made at the very beginning of the modeling that affects their implementation into practical—engineering problems. Such differences are

1. the required mathematical techniques that needed to customize an approach to a particular problem,
2. the number of the coefficients to be fitted,
3. the domain of applicability,
4. and the solution methods.

The last point will be discussed in details in Chap. 6. The second and third points are convenient to present together with the experiments. Regarding the first point, we refer only to the literature, and here, we are going to show only the simplest derivation as possible.

Maxwell–Cattaneo–Vernotte equation (MCV) The MCV equation is known as the first hyperbolic generalization of Fourier equation and extends the Fourier's law with the time derivative of the heat flux [54],

$$\tau \partial_t \mathbf{q} + \mathbf{q} = -\lambda \nabla T, \tag{56}$$

where τ is the so-called relaxation time and its presence expresses an inertia of heat conduction. Due to the hyperbolic property, the propagation speed remains finite and equal to $v_c = \sqrt{\alpha/\tau}$. Its derivation is based on the extension of variable space: $s = s(e, \mathbf{q})$ holds now. More precisely,

$$s(e, \mathbf{q}) = s_{eq}(e) - \frac{m}{2}\mathbf{q}^2, \tag{57}$$

where the first term expresses the classical local equilibrium assumption and its extension represents a deviation from this local equilibrium, that is, a phenomenon beyond the local equilibrium is considered. The quadratic extension of the specific entropy is the simplest choice in order to preserve its concavity properties, i.e., the thermodynamic stability. For further discussion, we refer to [55–57]. Equation (56) is used to explain a low-temperature phenomenon, called second sound. It is the damped waveform of heat conduction. Peshkov [58] observed it first in 1944 in superfluid He, which was predicted earlier by Tisza [59, 60] and Landau [61]. The precise mechanism behind the second sound is still not clear, however, later on, it has been found in solids, too, thanks to the theoretical work of Guyer and Krumhansl [62, 63].

In the framework of RET, a particular phonon picture is used for explanation in which there are two basic interaction processes among the phonons. The first one is a resistive process, characterized by its frequency or a time scale; and the overall energy is conserved. The second one is the so-called normal process when the momentum is also conserved and characterized by its frequency as well. The dominance of a particular collision interaction describes the way of heat conduction. In the case of resistive dominance, the describing model can be simplified to the Fourier equation, in the other way, the MCV equation may be considered. Assuming a particular mechanism also restricts the predicted propagation speed that was a subject of early researches. Recently, the MCV equation is tried to be found in room temperature experiments and still considered to be a valid extension of Fourier's law for heterogeneous materials, with moderate success.

Dual Phase Lag (DPL) and Jeffreys-type equations The DPL and Jeffreys-type equations are very similar to each other. Indeed, their linearized structure is the same. However, their thermodynamic origin is distinct. The DPL model is first proposed by Tzou [64], considering the Taylor series expansion of

$$\mathbf{q}(\mathbf{x}, t + \tau_q) = -\lambda \nabla T (\mathbf{x}, t + \tau_T), \tag{58}$$

where there are two different relaxation times τ_q and τ_T, each related to the heat flux and the temperature gradient, accordingly. This is not compatible with basic physical principles, constitutive equations cannot be introduced in this way [65]. It is criticized in recent papers [66–68] as well. Its thermodynamically compatible version is the Jeffreys-type model that can be derived easily, for instance, in the following way, using NET-IV. This is a convenient approach to extend the state space: introducing an internal variable that represents a short of internal degree of freedom [3, 4, 69]. Internal variables vanish in equilibrium state and one does not need to specify them with exact meaning at the beginning. It is possible to restrict only its tensorial order, for instance, let us assume that $s = s(e, \xi) = s_{eq}(e) - \frac{m}{2}\xi^2$ with ξ being a vectorial contribution to the entropy production. It is only customary to identify it as being the heat flux. The inequality of entropy production constraints the evolution of ξ, too, without knowing its exact role in the process. Since it is a vectorial variable, it must be coupled to the heat flux. Analogously to the previous calculations, the resulting system of equation is

$$q = l_{11} \nabla \frac{1}{T} + l_{12} \xi,$$

$$-m \partial_t \xi = l_{21} \nabla \frac{1}{T} + l_{22} \xi, \qquad (59)$$

when $J_s = \frac{q}{T}$. The conductivity coefficients are the parts of a matrix L that must be positive definite in order to fulfill the inequality. It means that $l_{11}, l_{22} \geq 0$ and $l_{11}l_{22} - l_{12}l_{21} > 0$. Then eliminating ξ, we obtain the DPL model:

$$\tau \partial_t q + q = -\lambda \nabla T - \hat{\tau} \partial_t (\nabla T), \qquad (60)$$

with the coefficients being

$$\tau = \frac{\rho m}{l_{22}}, \quad \lambda = \frac{1}{T^2} \left(l_{11} - \frac{l_{12}l_{21}}{l_{22}} \right), \quad \hat{\tau} = \frac{\rho m}{l_{22}} \frac{l_{11}}{T^2} = \tau \frac{l_{11}}{T^2}. \qquad (61)$$

It is important to emphasize that the coefficients τ and $\hat{\tau}$ are not independent of each other and could be 'compatible' until the first order Taylor series expansion of (58). Omitting that background, the outcome results in mathematical and physical shortcomings. It reduces to the MCV equation when $l_{11} = 0$ is considered. As a consequence, $l_{12}l_{21} < 0$, then the coupling becomes antisymmetric in order to preserve the positivity of the thermal conductivity λ. Moreover, it is not possible to obtain a time lag only for the temperature gradient and these properties become visible only using a strict thermodynamical derivation for the constitutive equations. Despite the shortcomings found in regard the background of DPL models, it is quite popular due to its relatively simple interpretation and the analogous treatment to the MCV equation [70–72].

Guyer–Krumhansl (GK) equation In the period 1960–70, it turned out that the MCV equation is not enough and cannot be the 'ultimate' or the last extension of the Fourier's law. The GK equation extends the MCV one with the Laplacian of the heat flux:

$$\tau \partial_t q + q = -\lambda \nabla T + \kappa^2 \Delta q, \qquad (62)$$

with κ being related to the mean free path of phonons or kind of a characteristic length scale of the process. Although Guyer and Krumhansl derived this model using a linearization of Boltzmann equation, there is a different way to obtain Eq. (62) without using any assumption about phonons [12]. Our set of variables are e and q, and now let us generalize the entropy current density as well:

$$J_s = B \cdot q, \qquad (63)$$

where B is a second order tensor, called current or Nyíri-multiplier [35]. Using this generalization of the entropy current density, it permits to obtain coupling between different tensorial order quantities for isotropic materials through the current multiplier. It can be called as *'entropic coupling'*. As a mathematical consequence of Eq. (63), the resulting model will always be parabolic and the solution of the entropy inequality yields the expression for $\mathbf{B}$. It is also possible to restrict the form of $\mathbf{B}$ using gradients of the non-equilibrium variables such as the heat flux [73–76]. There a particular form is used, for example

$$\mathbf{J}_s = \frac{\mathbf{q}}{T} + \mu \nabla \mathbf{q} \cdot \mathbf{q}. \tag{64}$$

It can be recovered when $\mathbf{B} = \frac{1}{T}\mathbf{I} + \mu \nabla \mathbf{q}$. Interestingly, the same form is proposed by the second law itself, the entropy inequality also restricts $\mathbf{B}$ without any further assumption. That is,

$$s(e, \mathbf{q}) = s_{eq}(e) - \frac{m}{2}\mathbf{q}^2, \quad \mathbf{J}_s = \mathbf{B} \cdot \mathbf{q}, \tag{65}$$

and then the entropy production is

$$\sigma_s = \nabla \mathbf{q} : \left(\mathbf{B} - \frac{1}{T}\mathbf{I} \right) + \mathbf{q} \cdot (\nabla \cdot \mathbf{B} - \rho m \partial_t \mathbf{q}) \geq 0, \tag{66}$$

which has a solution

$$\mathbf{B} - \frac{1}{T}\mathbf{I} = l_1 \nabla \mathbf{q},$$
$$\nabla \cdot \mathbf{B} - \rho m \partial_t \mathbf{q} = l_2 \mathbf{q}, \tag{67}$$

with $l_1 \geq 0$ and $l_2 \geq 0$, and considering $l_1 = \mu$ recovers the compatibility between (63) and (64). Moreover,

$$\tau = \frac{\rho m}{l_2}, \quad \lambda = \frac{1}{l_2 T^2}, \quad \kappa^2 = \frac{l_1}{l_2}. \tag{68}$$

Eliminating $\mathbf{B}$, the Eq. (62) is obtained, and can be reduced to the MCV equation taking $l_1 = 0$ in (67). The GK equation has a significant role in modeling heterogeneous materials such as rocks, foams, biological materials and other complicated processes [77–81]. Note that for such a generalized heat equation, the temperature gradient cannot be defined easily as a boundary condition. That problem is going to be discussed in Chap. 6.

Ballistic models In the 1960s, McNelly et al. [82–84] managed to measured experimentally the third way of heat conduction that called ballistic propagation. From an experimental point of view, it is measured as a heat wave propagating with the speed of sound. However, from a theoretical point of view, its exact interpretation is still not yet completely understood, depends on the particular approach used in the modeling process. That propagation speed reflects a thermo-mechanical coupling behind the phenomenon. Now let us show the difference between two possible approaches.

1. As previously mentioned, RET inherits the methodology from kinetic theory, where phonons are responsible for heat conduction phenomenon, with resistive and normal interactions. In the case of ballistic conduction, there is no interaction among them. Here, the scattering on the boundary has significance; it limits the free flow of phonons. Since the direct solution of the Boltzmann equation would be too difficult, an approximation is used, called momentum series expansion [85]. It results in a system with the size of infinite number of partial differential equations, having the following structure in 1+1 D:

$$\frac{\partial u_{\langle n \rangle}}{\partial t} + \frac{n^2}{4n^2 - 1} c \frac{\partial u_{\langle n-1 \rangle}}{\partial x} + c \frac{\partial u_{\langle n+1 \rangle}}{\partial x} = \begin{cases} 0 & n = 0 \\ -\frac{1}{\tau_R} u_{\langle 1 \rangle} & n = 1 \\ -\left(\frac{1}{\tau_R} + \frac{1}{\tau_N} \right) u_{\langle n \rangle} & 2 \leq n \leq N \end{cases}$$

(69)

Here, u denotes the corresponding quantity, such as u_0 is the energy density, u_1 is the momentum density and so on. For details, we refer to the work of Dreyer and Struchtrup [85]. Its disadvantage is that one must apply at least ca. 30 momentum equations in order to approximate the speed of sound, in one dimension. It also has to be noted here, that the momentum equations consist quantities with increasing tensorial order, i.e., a 30th order tensor in the 30th equation. In a general three dimensional problem, it means millions of equations, for quantities with unknown boundary conditions. Nevertheless, it was the first model which was able to include the ballistic contribution into the modeling and suggested the coupling between the heat flux and the pressure. That coupling realizes the connection between the thermal and mechanical fields at the level of constitutive equations. It would not be possible in the framework of CIT. Providing a wider picture about other approaches originating from kinetic theory, we also refer to [86–93].

2. Using the continuum approach of NET-IV (see also [94]), we can obtain the simplest model of ballistic propagation by extending the set of variables with a simple second-order internal variable $\mathbf{Q}$: $s = s(e, \mathbf{q}, \mathbf{Q})$,

$$s = s_{eq}(e) - \frac{m_1}{2} \mathbf{q}^2 - \frac{m_2}{2} \mathbf{Q}^2,$$

(70)

however, the entropy current density remains $\mathbf{J}_s = \mathbf{B}\mathbf{q}$. Comparing the one-dimensional form of the resulting system of equations

$$\tau_q \partial_t q + q + \lambda \partial_x T + \kappa \partial_x Q = 0,$$
$$\tau_Q \partial_t Q + Q + \kappa \partial_x q = 0,$$

(71)

to Eq. (69), the meaning of $\mathbf{Q}$ turns out to be the current density of the heat flux $\mathbf{q}$. When τ_Q is assumed to be zero, the system (71) reduces to the Guyer–Krumhansl equation. Within a continuum thermodynamical approach, the ballistic propagation is interpreted as an elastic wave. The essential difference is about the coefficients. In contrary to the RET model, the parameter κ is not fixed in (71) and can be used to adjust the experimentally measured propagation speed. Thus κ represents an extra freedom in the constitutive laws that allows to significantly reduce the number of equations, comparing to the RET, i.e., three equations are enough [95, 96].

The modeling of this particular phenomenon perfectly reflects the differences between various approaches. Also, in the case of the Guyer–Krumhansl equation, it matters how one derives the model. If one uses a kinetic approach, then the resulting model is seemingly restricted on that particular setting such as low-temperature heat conduction. On the contrary, the presented NET-IV derivation is independent of these assumptions. Despite that one obtains the same equation, the derivation itself disengage the limits and widen the validity region of the GK equation; hence it becomes possible to use such a model for room temperature problems, too.

In the following, we shall discuss further applications for mechanical models—the previously introduced approach of non-equilibrium thermodynamics with internal variables and current multipliers also applicable for those cases with the same steps as above. Here, we note that the extended diffusion models, related to the generalizations of Fick's law, have a common ground with heat conduction models, also possessing the same structure. Hence we omit their discussion, and we refer to [97].

2.2   Extensions of Navier–Stokes–Fourier System

In the case of heat conduction, it is more apparent how the structure of material influences the heat transport. In contrary, in a modeling problem of a rarefied gas, it is less apparent because the molecules remain the same as in its dense state. Furthermore, the modeling of collisions, how we describe the interaction among the molecules, is also the same. The most straightforward way to characterize the rarefaction state is using the so-called Knudsen number, which is the ratio of the mean free path and the size of the system. In a time-dependent process, it is possible to modify this definition and substitute the lengths with the corresponding characteristic time scales [98]. The mean free path that describes the average distance that particles travel between two collisions depends on the temperature and the pressure, in other words, the mass density ρ and the temperature T characterizes the rarefaction state of a gas. According to the experimental background of rarefied gases, the classical system of Navier–Stokes–Fourier equations loses its validity around Kn $\approx 0.05 -$ 0.1 [10]. For the detailed kinetic modeling of rarefied gases, we refer to the work of Struchtrup [10, 98].

In agreement of the previous statement, the mass density ρ is in the state space, i.e., $s = s_{eq}(e, \rho) - \frac{m_1}{2}\mathbf{q}^2 - \frac{m_2}{2}\mathbf{\Pi}_{\text{dev}}^2 - \frac{m_3}{6}\mathbf{\Pi}_{\text{sph}}^2$ in which the tensorial internal variable $\mathbf{Q}$ is identified as being the dynamic pressure $\mathbf{\Pi}$. This is the usual approach of EIT [74]. In a rarefied state, the compressibility becomes essential, and its proper thermodynamic modeling demands the deviatorical-spherical decomposition at this level. Since these parts are linearly independent of each other, they stand as independent variables in the state space and possess a separate evolution equation with distinct relaxation times. Both will be coupled to the heat flux. Moreover, we also apply this decomposition on the current multiplier $\mathbf{B}$ in the entropy current density as $\mathbf{J}_s = (\mathbf{B}_{\mathrm{dev}} + \frac{1}{3}B_{\mathrm{sph}}\mathbf{I})\mathbf{q}$. Therefore, the spherical parts of pressure and current multiplier will be coupled in the constitutive equations:

$$
\begin{aligned}
\tau_1 \partial_t \mathbf{q} + \mathbf{q} &= -\lambda \nabla T + \alpha_{21}\nabla \Pi_{\mathrm{sph}} + \beta_{21}\nabla \cdot \boldsymbol{\Pi}_{\mathrm{dev}}, \\
\tau_2 \partial_t \boldsymbol{\Pi}_{\mathrm{dev}} + \boldsymbol{\Pi}_{\mathrm{dev}} &= -\nu(\nabla \mathbf{v})_{\mathrm{dev}} - \beta_{12}(\nabla \mathbf{q})_{\mathrm{dev}} \\
\tau_3 \partial_t \Pi_{\mathrm{sph}} + \Pi_{\mathrm{sph}} &= -\eta \nabla \cdot \mathbf{v} - \alpha_{12}\nabla \cdot \mathbf{q},
\end{aligned}
\tag{72}
$$

where the τ_i ($i = 1, 2, 3$) are the corresponding relaxation times, ν and η are the shear and bulk viscosities, respectively, and the α, β parameters are the coupling coefficients [42]. We make the following remarks:

1. Its counterpart in RET, derived by Arima et al. [33, 99], is compatible with Eq. (72) as having the same structure and the coefficients make the essential difference. While in the RET based model, the α and β coupling coefficients are fixed, together with the transport coefficients (λ, ν and η):

$$
\lambda = (1 + c_v^*)R^2 \rho_0 T_0 \tau_1, \quad \nu = 2R\rho_0 T_0 \tau_2, \quad \eta = \left(\frac{2}{3} - \frac{1}{c_v^*}\right)R\rho_0 T_0 \tau_3,
$$

$$
\alpha_{12} = \frac{2c_v^* - 3}{3c_v^*(1 + c_v^*)}\tau_3, \quad \alpha_{21} = RT_0\tau_1, \quad \beta_{12} = \frac{2}{1 + c_v^*}\tau_2, \quad \beta_{21} = RT_0\tau_1,
$$

with $c_v^* = c_v/R$. Furthermore, the calculated (theoretical) transport coefficients (λ, ν and η) are independent of the mass density since the relaxation times are inversely proportional with ρ [100]. In contrary, there are several measurement in the literature that show some density dependence both in the dense and rarefied regions [101–106]. From this point of view, it is possible to distinguish 'effective' (measurable) and 'physical' (theoretical) transport coefficients.
   In a continuum model, such as the NET-IV, the measured viscosities and thermal conductivity are used, that is, it makes no preposition about the theoretical values of the transport coefficients. It significantly influences the applicability of a model which becomes apparent in the evaluation of experiments.

2. It is visible that Eq. (72) has the same structure as the ballistic heat conduction models previously. If $\tau_2 = \tau_3 = 0$ are considered, then it reduces to a Guyer–Krumhansl-type equation for heat conduction. Moreover, it also shares the freedom to adjust the coupling coefficients, in Eq. (72) those are free to fit.

3. In this case, the internal variables are specified as being the heat flux and pressure. That approach would be more general without this specification.

4. Equation (72) is the simplest applicable extension of the Navier–Stokes–Fourier system. The coupling arises naturally from the entropy production as a requirement. Thus its immediate consequence is that the heat flux could have a significant

contribution to the pressure field and vice versa, which reflects the present ballistic effects.

5. Since the same generalization of the specific entropy and its current density is applied, it can be called as a 'non-local generalization of the specific entropy'.

2.3   Extensions of Hooke's Law: Rheology

Rheology, i.e., creeping and stress relaxation occurs in many engineering problems, and expresses the limits of the Hooke's law, especially in time-dependent situations. These effects mainly depend on the temperature conditions and the material parameters. The relevant time scale in which such mechanical effect occurs is also a function of the material properties, and, in parallel, the boundary conditions.

In contrary to the generalized heat equations, the simplest rheological models possess only memory extensions without any non-local couplings to other quantities which is possible only with the classical entropy current density $\mathbf{q}/T$. The well-know rheological equations can be derived on the same basis using the approach of NET-IV, assuming a single tensorial internal variable $\boldsymbol{\xi}$, that is, $s = s(e, \varepsilon, \boldsymbol{\xi})$, without specifying the exact meaning of $\boldsymbol{\xi}$ and

$$s(e, \varepsilon, \boldsymbol{\xi}) = s_{\text{eq}}(e, \varepsilon) - \frac{1}{2}\text{Tr}(\boldsymbol{\xi}^2), \tag{73}$$

in analogy with the previous models. More importantly, the rheological effects reflect the non-equilibrium mechanical behavior of a material, hence the stress $\boldsymbol{\sigma}$ acquire an extra term: $\boldsymbol{\sigma} = \tilde{\boldsymbol{\sigma}} + \hat{\boldsymbol{\sigma}}$ with $\boldsymbol{\sigma}$ being the total stress and $\hat{\boldsymbol{\sigma}}$ expresses the rheological contribution besides the pure elastic part $\tilde{\boldsymbol{\sigma}}$. Consequently, it results in an additional term in the mechanical source of internal energy balance,

$$\rho D_t e + \nabla \cdot \mathbf{q} = \text{Tr}(\tilde{\boldsymbol{\sigma}} d_t \varepsilon) + \text{Tr}(\hat{\boldsymbol{\sigma}} d_t \varepsilon). \tag{74}$$

For simplicity, let us restrict ourselves to the one-dimensional situation in which the entropy production yields a simple inequality

$$\hat{\sigma} d_t \varepsilon - \rho T \xi d_t \xi \geq 0, \tag{75}$$

which can be solved in the same way using Onsager's approach,[8]

$$\hat{\sigma} = l_{11} d_t \varepsilon + l_{12}(-\rho T \xi),$$
$$d_t \xi = l_{21} d_t \varepsilon + l_{22}(-\rho T \xi) \tag{76}$$

---

[8]We note that this identification of fluxes and forces is not unique, and it affects the definitions of material parameters.

with $l_{11}, l_{22} \geq 0$ and $\det \mathbf{L}^{\text{sym}} \geq 0$ where $\mathbf{L}$ consists the l_{ij} conduction coefficients and the second law restricts its symmetric part only as the antisymmetric part does not contribute to the entropy production. After eliminating the internal variable ξ, we arrive to the Kluitenberg–Verhás model

$$\tau d_t \sigma + \sigma = E_0 \varepsilon + E_1 d_t \varepsilon + E_2 d_t^2 \varepsilon, \tag{77}$$

with $\tau, E_1 > 0$ and $E_0, E_2 \geq 0$. The Kluitenberg–Verhás model (77) can be reduced to the following special cases,

1. Hooke model: $\sigma = E_0 \varepsilon$,
2. Kelvin–Voigt model: $\sigma = E_0 \varepsilon + E_1 d_t \varepsilon$,
3. Maxwell model (fluids): $\tau d_t \sigma + \sigma = E_1 d_t \varepsilon$,
4. Poynting–Thomson–Zener model: $\tau d_t \sigma + \sigma = E_0 \varepsilon + E_1 d_t \varepsilon$,
5. Jeffreys model (fluids): $\tau d_t \sigma + \sigma = E_1 d_t \varepsilon + E_2 d_t^2 \varepsilon$,

for the detailed derivation and their three-dimensional formulation, we refer to [22, 107, 108]. Finally, the second law allows the following non-trivial combination of the parameters as well:

$$I_1 = E_1 - \tau E_0 \geq 0, \quad I_2 = E_2 - \tau I_1 \lessgtr 0. \tag{78}$$

The parameter I_1 is called index of damping and I_2 being the index of inertia. The above relations exclude a few models on the thermodynamical ground, i.e., otherwise, the second law would be violated. On the other hand, the case of $I_2 > 0$ includes the possibility of local rheological resonance that is an entirely different one than the one which called elastic resonance where the elastic waves are in tune with the excitation [22].

From a practical point of view, the rheological modeling is not that straightforward. Besides the difficulties in the experiments—for instance: uniaxiality, temperature effects, determining and maintaining the boundary conditions [21, 24, 109]—the theoretical background is more diverse and can depend on the material type in question. For example, the book of Mewis and Wagner [110] deals with colloidal suspensions by introducing power-law models. The book of Chhabra and Richardson [111] gives a better insight into the engineering aspects, independently of the material structure. We also note here an interesting future application of rock's rheology regarding the next generation gravitation wave detector, called Einstein Telescope [112]. Since it is planned to be an underground detector, it becomes vital to filter the corresponding noise sources that arise due to the rheological behavior [113, 114].

References

1. T. Matolcsi, *Ordinary Thermodynamics* (Akadémiai Kiadó, Budapest, 2004)
2. I. Gyarmati, *Non-Equilibrium Thermodynamics* (Springer, Berlin, 1970)

3. A. Berezovski, P. Ván, *Internal Variables in Thermoelasticity* (Springer, Berlin, 2017)
4. J. Verhás, *Thermodynamics and Rheology* (Akadémiai Kiadó-Kluwer Academic Publisher, Dordrecht, 1997)
5. S.R. de Groot, P. Mazur, *Non-Equilibrium Thermodynamics* (Dover Publications, Mineola, 1963)
6. D. Jou, J. Casas-Vázquez, G. Lebon, *Extended Irreversible Thermodynamics*, 4th edn. (Springer, New York, 2010)
7. I. Müller, T. Ruggeri, *Rational Extended Thermodynamics* (Springer, Berlin, 1998)
8. D. Jou, J. Casas-Vázquez, G. Lebon, Extended irreversible thermodynamics. Rep. Prog. Phys. **51**(8), 1105 (1988)
9. G. Lebon, D. Jou, J. Casas-Vázquez, *Understanding Non-equilibrium Thermodynamics* (Springer, Berlin, 2008)
10. H. Struchtrup, *Macroscopic Transport Equations for Rarefied Gas Flows* (Springer, Berlin, 2005)
11. T. Matolcsi, P. Ván, Can material time derivative be objective? Phys. Lett. A **353**(2), 109–112 (2006)
12. R. Kovács, P. Ván, Generalized heat conduction in heat pulse experiments. Int. J. Heat Mass Transf. **83**, 613–620 (2015)
13. G. Vayakis, T. Sugie, T. Kondoh, T. Nishitani, E. Ishitsuka, M. Yamauchi, H. Kawashima, T. Shikama, Radiation-induced thermoelectric sensitivity (RITES) in ITER prototype magnetic sensors. Rev. Sci. Instrum. **75**(10), 4324–4327 (2004)
14. A. Loarte, G. Saibene, R. Sartori, V. Riccardo, P. Andrew, J. Paley, W. Fundamenski, T. Eich, A. Herrmann, G. Pautasso et al., Transient heat loads in current fusion experiments, extrapolation to ITER and consequences for its operation. Phys. Scr. **2007**(T128), 222 (2007)
15. B. Smith, P.P.H. Wilson, M.E. Sawan, Three dimensional neutronics analysis of the ITER first wall/shield module 13, in *2007 IEEE 22nd Symposium on Fusion Engineering* (IEEE, 2007), pp. 1–4
16. A. Huber, A. Arakcheev, G. Sergienko, I. Steudel, M. Wirtz, A.V. Burdakov, J.W. Coenen, A. Kreter, J. Linke, Ph Mertens et al., Investigation of the impact of transient heat loads applied by laser irradiation on iter-grade tungsten. Phys. Scr. **2014**(T159), 014005 (2014)
17. Yi Zhu, Liu Hong, Zaibao Yang, Wen-An Yong, Conservation-dissipation formalism of irreversible thermodynamics. J. Non-Equilib. Thermodyn. **40**(2), 67–74 (2015)
18. Lars Onsager, Reciprocal relations in irreversible processes. I. Phys. Rev. **37**(4), 405 (1931)
19. Lars Onsager, Reciprocal relations in irreversible processes. II. Phys. Rev. **38**(12), 2265 (1931)
20. V.A. Cimmelli, D. Jou, T. Ruggeri, P. Ván, Entropy principle and recent results in non-equilibrium theories. Entropy **16**(3), 1756–1807 (2014)
21. T. Fülöp, P. Ván, A. Csatár, Elasticity, plasticity, rheology and thermal stres–an irreversible thermodynamical theory. Elastic **2**(7) (2013)
22. T. Fülöp, Cs. Asszonyi, P. Ván, Distinguished rheological models in the framework of a thermodynamical internal variable theory. Contin. Mech. Thermodyn. **27**(6), 971–986 (2015)
23. T. Fülöp, Objective Thermomechanics (2015). arXiv:1510.08038
24. C. Asszonyi, A. Csatár, T. Fülöp, Elastic, thermal expansion, plastic and rheological processes-theory and experiment (2015). arXiv:1512.05863
25. T. Fülöp, R. Kovács, Á. Lovas, Á. Rieth, T. Fodor, M. Szücs, P. Ván, Gy. Gróf, Emergence of non-Fourier hierarchies. Entropy **20**(11), 832 (2018). arXiv:1808.06858
26. S. Kjelstrup, D. Bedeaux, *Non-Equilibrium Thermodynamics of Heterogeneous Systems*, vol. 16 (World Scientific, Singapore, 2008)
27. P. Ván, R. Kovács, Variational principles and thermodynamics (2019). Submitted arxiv:1908.02679
28. G. Fichera, Is the Fourier theory of heat propagation paradoxical? Rendiconti del Circolo Matematico di Palermo **41**(1), 5–28 (1992)
29. G. Lebon, From classical irreversible thermodynamics to extended thermodynamics. Acta Phys. Hung. **66**(1–4), 241–249 (1989)

30. D. Jou, J. Casas-Vazquez, G. Lebon, Extended irreversible thermodynamics revisited (1988–98). Rep. Prog. Phys. **62**(7), 1035 (1999)
31. G. Lebon, Heat conduction at micro and nanoscales: a review through the prism of extended irreversible thermodynamics. J. Non-Equilib. Thermodyn. **39**(1), 35–59 (2014)
32. V. Ciancio, L. Restuccia, On heat equation in the framework of classic irreversible thermodynamics with internal variables. Int. J. Geom. Methods Mod. Phys. **13**(08), 1640003 (2016)
33. T. Ruggeri, M. Sugiyama, *Rational Extended Thermodynamics Beyond the Monatomic Gas* (Springer, Berlin, 2015)
34. B. Nyíri. On the extension of the governing principle of dissipative processes to nonlinear constitutive equations. Acta Phys. Hung. **66**(1), 19–28 (1989)
35. B. Nyíri, On the entropy current. J. Non-Equilib. Thermodyn. **16**(2), 179–186 (1991)
36. V. Ciancio, V.A. Cimmelli, P. Ván, On the evolution of higher order fluxes in non-equilibrium thermodynamics. Math. Comput. Model. **45**, 126–136 (2007). arXiv:cond-mat/0407530
37. V.A. Cimmelli, Different thermodynamic theories and different heat conduction laws. J. Non-Equilib. Thermodyn. **34**(4), 299–333 (2009)
38. V.A. Cimmelli, P. Ván, The effects of nonlocality on the evolution of higher order fluxes in nonequilibrium thermodynamics. J. Math. Phys. **46**(11), 112901 (2005)
39. A. Sellitto, V.A. Cimmelli, D. Jou, *Mesoscopic Theories of Heat Transport in Nanosystems*, vol. 6 (Springer, Berlin, 2016)
40. D. Jou, V.A. Cimmelli, Constitutive equations for heat conduction in nanosystems and non-equilibrium processes: an overview. Commun. Appl. Ind. Math. **7**(2), 196–222 (2016)
41. A. Sellitto, V.A. Cimmelli, D. Jou, Nonequilibrium thermodynamics and heat transport at nanoscale, in *Mesoscopic Theories of Heat Transport in Nanosystems* (Springer International Publishing, Berlin, 2016), pp. 1–30
42. R. Kovács, D. Madjarevic, S. Simic, P. Ván, Theories of rarefied gases (2018). Submitted arXiv:1812.10355
43. P. Rogolino, R. Kovács, P. Ván, V.A. Cimmelli, Generalized heat-transport equations: parabolic and hyperbolic models. Contin. Mech. Thermodyn. **30**, AiP–14 (2018)
44. H.C. Öttinger, *Beyond Equilibrium Thermodynamics* (Wiley, Berlin, 2005)
45. M. Grmela, H.C. Öttinger, Dynamics and thermodynamics of complex fluids. I. Development of a general formalism. Phys. Rev. E **56**(6), 6620 (1997)
46. H.C. Öttinger, M. Grmela, Dynamics and thermodynamics of complex fluids. II. Illustrations of a general formalism. Phys. Rev. E **56**(6), 6633 (1997)
47. M. Pavelka, V. Klika, M. Grmela, *Multiscale Thermo-Dynamics: Introduction to GENERIC* (Walter de Gruyter GmbH & Co KG, Berlin, 2018)
48. M. Grmela, M. Pavelka, V. Klika, B.-Y. Cao, N. Bendian, Entropy and entropy production in multiscale dynamics. J. Non-Equilib. Thermodyn. (2019). Online first
49. M. Grmela, L. Restuccia, Nonequilibrium temperature in the multiscale dynamics and thermodynamics. Atti della Accademia Peloritana dei Pericolanti-Classe di Scienze Fisiche, Matematiche e Naturali **97**(S1), 8 (2019)
50. P.C. Hohenberg, B.I. Halperin, Theory of dynamic critical phenomena. Rev. Mod. Phys. **49**(3), 435–479 (1977)
51. O. Penrose, P.C. Fife, On the relation between the standard phase-field model and a "thermodynamically consistent" phase-field model. Phys. D **69**, 107–113 (1993)
52. L.-Q. Chen, Phase-field models for microstructure evolution. Annu. Rev. Mater. Res. **32**(1), 113–140 (2002)
53. W.J. Boettinger, J.A. Warren, C. Beckermann, A. Karma, Phase-field simulation of solidification. Annu. Rev. Mater. Res. **32**(1), 163–194 (2002)
54. I. Gyarmati, On the wave approach of thermodynamics and some problems of non-linear theories. J. Non-Equilib. Thermodyn. **2**, 233–260 (1977)
55. A. Sellitto, V. Tibullo, Y. Dong, Nonlinear heat-transport equation beyond Fourier law: application to heat-wave propagation in isotropic thin layers. Contin. Mech. Thermodyn. **29**(2), 411–428 (2017)

56. Hai-Dong Wang, Bing-Yang Cao, Zeng-Yuan Guo, Non-Fourier heat conduction in carbon nanotubes. J. Heat Transf. **134**(5), 051004 (2012)
57. Ben-Dian Nie, Bing-Yang Cao, Three mathematical representations and an improved ADI method for hyperbolic heat conduction. Int. J. Heat Mass Transf. **135**, 974–984 (2019)
58. V. Peshkov, Second sound in helium II. J. Phys. (Moscow) **381**(8) (1944)
59. L. Tisza, Transport phenomena in Helium II. Nature **141**, 913 (1938)
60. L. Tisza, The theory of liquid Helium. Phys. Rev. **72**(9), 838–877 (1947)
61. L. Landau, On the theory of superfluidity of Helium II. J. Phys. **11**(1), 91–92 (1947)
62. R.A. Guyer, J.A. Krumhansl, Solution of the linearized phonon Boltzmann equation. Phys. Rev. **148**(2), 766–778 (1966)
63. R.A. Guyer, J.A. Krumhansl, Thermal conductivity, second sound and phonon hydrodynamic phenomena in nonmetallic crystals. Phys. Rev. **148**(2), 778–788 (1966)
64. D.Y. Tzou, *Macro- to Micro-scale Heat Transfer: The Lagging Behavior* (CRC Press, Boca Raton, 1996)
65. R. Kovács, P. Ván, Thermodynamical consistency of the Dual Phase Lag heat conduction equation. Contin. Mech. Thermodyn. 1–8 (2017)
66. M. Fabrizio, B. Lazzari, V. Tibullo, Stability and thermodynamic restrictions for a dual-phase-lag thermal model. J. Non-Equilib. Thermodyn. (2017)
67. S.A. Rukolaine, Unphysical effects of the dual-phase-lag model of heat conduction. Int. J. Heat Mass Transf. **78**, 58–63 (2014)
68. S.A. Rukolaine, Unphysical effects of the dual-phase-lag model of heat conduction: higher-order approximations. Int. J. Therm. Sci. **113**, 83–88 (2017)
69. P. Ván, A. Berezovski, J. Engelbrecht, Internal variables and dynamic degrees of freedom. J. Non-Equilib. Thermodyn. **33**(3), 235–254 (2008)
70. C. D'Apice, S. Chirita, V. Zampoli, On the well-posedness of the time-differential three-phase-lag thermoelasticity model. Arch. Mech. **68**(5) (2016)
71. S. Chirita, V. Zampoli, Spatial behavior of the dual-phase-lag deformable conductors. J. Therm. Stress. **41**(10–12), 1276–1296 (2018)
72. V. Zampoli, Uniqueness theorems about high-order time differential thermoelastic models. Ricerche di Matematica **67**(2), 929–950 (2018)
73. D. Jou, C. Perez-Garcia, L.S. Garcia-Colin, M.L. De Haro, R.F. Rodriguez, Generalized hydrodynamics and extended irreversible thermodynamics. Phys. Rev. A **31**(4), 2502 (1985)
74. G. Lebon, A. Cloot, Propagation of ultrasonic sound waves in dissipative dilute gases and extended irreversible thermodynamics. Wave Motion **11**, 23–32 (1989)
75. A. Sellitto, V.A. Cimmelli, D. Jou, Entropy flux and anomalous axial heat transport at the nanoscale. Phys. Rev. B (87), 054302 (2013)
76. V. Ciancio, L. Restuccia, A derivation of a Guyer-Krumhansl type temperature equation in classical irreversible thermodynamics with internal variables. Atti della Accademia Peloritana dei Pericolanti-Classe di Scienze Fisiche, Matematiche e Naturali **97**(S1), 5 (2019)
77. S. Both, B. Czél, T. Fülöp, Gy. Gróf, Á. Gyenis, R. Kovács, P. Ván, J. Verhás, Deviation from the Fourier law in room-temperature heat pulse experiments. J. Non-Equilib. Thermodyn. **41**(1), 41–48 (2016)
78. P. Ván, A. Berezovski, T. Fülöp, Gy. Gróf, R. Kovács, Á. Lovas, J. Verhás. Guyer-Krumhansl-type heat conduction at room temperature. EPL **118**(5), 50005 (2017). arXiv:1704.00341v1
79. M. Calvo-Schwarzwälder, T.G. Myers, M.G. Hennessy, The one-dimensional Stefan problem with non-Fourier heat conduction (2019). arXiv:1905.06320
80. M.G. Hennessy, M. Calvo-Schwarzwälder, T.G. Myers, Modelling ultra-fast nanoparticle melting with the Maxwell-Cattaneo equation. Appl. Math. Model. **69**, 201–222 (2019)
81. M.G. Hennessy, M.C. Schwarzwälder, T.G. Myers, Asymptotic analysis of the Guyer-Krumhansl-Stefan model for nanoscale solidification. Appl. Math. Model. **61**, 1–17 (2018)
82. T.F. McNelly, Second sound and anharmonic processes in isotopically pure Alkali-Halides. Ph.D. Thesis, Cornell University (1974)
83. H.E. Jackson, C.T. Walker, T.F. McNelly, Second sound in NaF. Phys. Rev. Lett. **25**(1), 26–28 (1970)

84. H.E. Jackson, C.T. Walker, Thermal conductivity, second sound and phonon-phonon interactions in NaF. Phys. Rev. B **3**(4), 1428–1439 (1971)
85. W. Dreyer, H. Struchtrup, Heat pulse experiments revisited. Contin. Mech. Thermodyn. **5**, 3–50 (1993)
86. T.F. McNelly, S.J. Rogers, D.J. Channin, R.J. Rollefson, W.M. Goubau, G.E. Schmidt, J.A. Krumhansl, R.O. Pohl, Heat pulses in NaF: onset of second sound. Phys. Rev. Lett. **24**(3), 100–102 (1970)
87. V. Narayanamurti, R.C. Dynes, Ballistic phonons and the transition to second sound in solid ^{3}He and ^{4}He. Phys. Rev. B **12**(5), 1731–1738 (1975)
88. F.X. Alvarez, D. Jou, Memory and nonlocal effects in heat transport: from diffusive to ballistic regimes. Appl. Phys. Lett. **90**(8), 083109 (2007)
89. Y. Ma, A transient ballistic-diffusive heat conduction model for heat pulse propagation in nonmetallic crystals. Int. J. Heat Mass Transf. **66**, 592–602 (2013)
90. Y. Ma, A hybrid phonon gas model for transient ballistic-diffusive heat transport. J. Heat Transf. **135**(4), 044501 (2013)
91. Yu-Chao Hua, Bing-Yang Cao, Slip boundary conditions in ballistic-diffusive heat transport in nanostructures. Nanoscale Microscale Thermophys. Eng. **21**(3), 159–176 (2017)
92. Dao-Sheng Tang, Bing-Yang Cao, Superballistic characteristics in transient phonon ballistic-diffusive transport. Appl. Phys. Lett. **111**(11), 113109 (2017)
93. Han-Ling Li, Bing-Yang Cao, Radial ballistic-diffusive heat conduction in nanoscale. Nanoscale Microscale Thermophys. Eng. **23**(1), 10–24 (2019)
94. A. Famá, L. Restuccia, P. Ván, Generalized ballistic-conductive heat conduction in isotropic materials (2019). arXiv:1902.10980
95. R. Kovács, P. Ván, Models of Ballistic propagation of heat at low temperatures. Int. J. Thermophys. **37**(9), 95 (2016)
96. R. Kovács, P. Ván, Second sound and ballistic heat conduction: NaF experiments revisited. Int. J. Heat Mass Transf. **117**, 682–690 (2018). arXiv:1708.09770
97. D. Jou, J. Camacho, M. Grmela, On the nonequilibrium thermodynamics of non-Fickian diffusion. Macromolecules **24**(12), 3597–3602 (1991)
98. H. Struchtrup, Resonance in rarefied gases. Contin. Mech. Thermodyn. **24**(4–6), 361–376 (2012)
99. T. Arima, S. Taniguchi, T. Ruggeri, M. Sugiyama, Extended thermodynamics of dense gases. Contin. Mech. Thermodyn. **24**(4–6), 271–292 (2012)
100. V.K. Michalis, A.N. Kalarakis, E.D. Skouras, V.N. Burganos, Rarefaction effects on gas viscosity in the Knudsen transition regime. Microfluid. Nanofluidics **9**(4–5), 847–853 (2010)
101. A. Van Itterbeek, A. Claes, Measurements on the viscosity of hydrogen-and deuterium gas between 293 K and 14 K. Physica **5**(10), 938–944 (1938)
102. A. Van Itterbeek, W.H. Keesom, Measurements on the viscosity of helium gas between 293 and 1.6 k. Physica **5**(4), 257–269 (1938)
103. A. Van Itterbeek, O. Van Paemel, Measurements on the viscosity of gases for low pressures at room temperature and at low temperatures. Physica **7**(3), 273–283 (1940)
104. A. Michels, A.C.J. Schipper, W.H. Rintoul, The viscosity of hydrogen and deuterium at pressures up to 2000 atmospheres. Physica **19**(1–12), 1011–1028 (1953)
105. J.A. Gracki, G.P. Flynn, J. Ross, Viscosity of nitrogen, helium, hydrogen, and argon from −100 to 25 c up to 150–250 atmospheres. Project SQUID Technical Report (1969), p. 33
106. J.A. Gracki, G.P. Flynn, J. Ross, Viscosity of nitrogen, helium, hydrogen, and argon from −100 to 25 c up to 150–250 atm. J. Chem. Phys. **9**, 3856–3863 (1969)
107. T. Fülöp, M. Szücs, Analytical solution method for rheological problems of solids (2018). arXiv:1810.06350
108. M. Szücs, T. Fülöp, Kluitenberg-Verhás rheology of solids in the GENERIC framework. J. Non-Equilib. Thermodyn. **44**(3), 247–259 (2019). arXiv:1812.07052
109. L. Écsi, P. Ván, T. Fülöp, B. Fekete, P. Élesztős, R. Janco, A thermoelastoplastic material model for finite-strain cyclic plasticity of metals (2017). arXiv:1709.05416

110. J. Mewis, N.J. Wagner, *Colloidal Suspension Rheology* (Cambridge University Press, Cambridge, 2012)
111. R.P. Chhabra, J.F. Richardson, *Non-Newtonian Flow and Applied Rheology: Engineering Applications* (Butterworth-Heinemann, Oxford, 2011)
112. ET Science Team. Einstein gravitational wave telescope, conceptual design study. Technical Report ET-0106C-10 (2011). http://www.et-gw.eu/etdsdocument
113. G.G. Barnaföldi, T. Bulik, M. Cieslar, E. Dávid, M. Dobróka, E. Fenyvesi, D. Gondek-Rosinska, Z. Gráczer, G. Hamar, G. Huba et al., First report of long term measurements of the MGGL laboratory in the Mátra mountain range. Class. Quantum Gravity **34**(11), 114001 (2017)
114. P. Ván, G.G. Barnaföldi, T. Bulik, T. Biró, S. Czellár, M. Cieślar, Cs. Czanik, E. Dávid, E. Debreceni, M. Denys et al., Long term measurements from the Mátra gravitational and geophysical laboratory (2018). arXiv:1811.05198

Chapter 3
Applications in Renewable Energy

1 Sustainable Power Generation

> *When people talk about traveling to the past, they worry about changing the present by doing small things, but pretty much no one in the present thinks that they can change the future by doing something small.*[1]

The consequences of climate change are affecting our everyday life with increasing presence around us [1–4]. Hence, wise resource management is inevitable since our current consumer habits endanger humanity in the long term. The increasing level of oceans [5], the flooding rivers, more frequent wildland fires [6], more intense tropical cyclones [7] are causing continuously increasing problems. If the fossil fuel utilization is summed up with the caused losses [8], it becomes apparent that a dramatic change is required in the energy industry to mitigate the global warming [9] and survive our ever-increasing hunger for energy that is our challenge for this century. The other solution, i.e., using geoengineering tools [10], might cause even larger problems since they were not tested before, and there are only simulation results available.

Following the motivation above, a toolbox is prepared for engineers to get a glimpse to renewable energy technologies. This chapter starts with the discussion of solar and wind energy utilization. Even though there are other renewable technologies present like tidal power plants and water turbines, the thermal engineering approach of these solutions correlates well with that of wind turbines since mechanical movement is used for energy generation. Then combustion is discussed since several manufacturing processes require concentrated thermal power besides domestic heating and power generation. It is apparent that the number of combustion devices has to drop drastically in the upcoming decades, however, they will remain in use in several industries.

---

[1] *Source* reddit.com/r/Showerthoughts, 05/11/2016.

© Springer Nature Switzerland AG 2020
V. Józsa and R. Kovács, *Solving Problems in Thermal Engineering*, Power Systems,
https://doi.org/10.1007/978-3-030-33475-8_3

The focus in the following years will be on the utilization of otherwise wasted heat. Consequently, devices utilizing few tens to few hundreds °C temperature difference already gain an increased attention. Their efficiency ranges between 5 and 20%, nevertheless, it is significantly more than zero. Therefore, the building blocks of thermal cycles are discussed lastly to allow the reader to design a new device which revolutionizes the energy industry.

2 Solar Energy Utilization

Solar energy utilization in industrial scales went through a rapid development since the millennium [11–13], and the trend goes on. The price of a solar cell is continuously decreasing as the demand for them is increasingly growing [14–16]. These effects together lead to positive influence on both economy and sustainability [17] and decreasing the carbon footprint of its production [18–20]. Solar collectors principally generate heat energy which can be used either for hot water production or steam generation for industrial processes [21] or driving a steam turbine to generate electricity [22]. Even though Concentrated Solar Power technology can be used for electricity generation with higher efficiency than photovoltaic panels, the more complicated system is less popular among investors [23]. The obvious advantage of solar collectors is that the heat transfer medium can be stored, hence, a seamless and predictable operation can be ensured.

Therefore, the present section highlights solar collector technology at first since it is a good platform to start with thermal system analyses. Then solar cells are discussed which development is driven by surface and material technologies. There are alternative solar energy utilization solutions like the solar chimney [24, 25], however, they are omitted now due to their low efficiency compared to the two flagship technologies.

2.1 Solar Collectors

The principal existing and emerging concentrated solar power technologies are shown in Fig. 1. The difference between them is how the heat of solar radiation is concentrated on the working fluid (if it performs work, e.g., drives a steam turbine) or on the heat transfer fluid (if only its heat is utilized, and the medium does not perform work in the thermodynamical sense). While Parabolic Trough Collectors (PTC) are complete units which are easily scalable [11, 26, 27], Solar Towers can be used as complete units and they usually associated with high thermal/electric power [28–31]. There were gaps in small-scale concentrated and scalable concentrated solutions which were filled by the Linear Fresnel Reflector [32–34] and the Parabolic Dish [35–37]. Note that the cited references after the mentioned technologies provide a good starting point for deeper details on design and technology advancement. Either

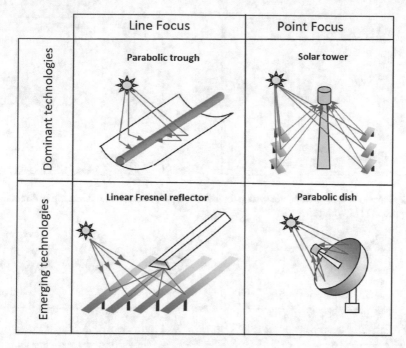

Fig. 1 Notable existing and emerging concentrated solar collector technologies [29]. With permission

industrial or domestic solar collector type is used, their surfaces need frequent cleaning to maintain their high operational efficiency [38].

Figure 2 shows the use of PTCs for steam generation. Pranesh et al. [21] summarized several technological processes in the temperature range of 60–260 °C where this technology can be efficiently used. Therefore, various industries may exploit the advantage of these systems. For continuous operation, a thermal storage tank is necessary that can be easily solved in the double-loop circuit. Hence, the disadvantage of the direct steam generation is its sensitivity to the variation of the solar flux during the day.

The steam generation is a well-controlled process in boilers. However, the problem of dry-out causes severe damage to the system [39, 40], hence, the steam in the drop tube always should be in the wet regime. PTCs generate steam in the same way, the phases of water evaporation in a tubular flow is shown in Fig. 3. The difficulty arises when the solar flux fluctuates which is otherwise a natural phenomenon. Therefore, direct steam generation is hard to implement into most practical systems.

The optimization procedure of a Central Receiver System solar collector is shown in Fig. 4. The first step is the determination of the optimal solar flux distribution [41–43], then the field layout has to be optimized [44–46].

The following two types are domestic solar collectors. They feature no moving parts to track the movement of the Sun, and they can be easily installed on rooftops.

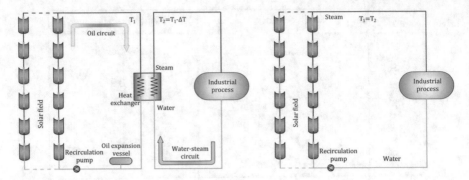

Fig. 2 Parabolic Trough Collector system for industrial processes. Left: indirect, right: direct steam generation [11]. With permission

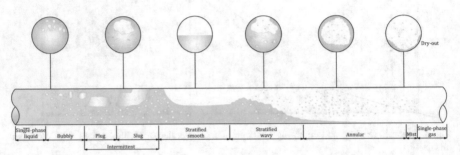

Fig. 3 Boiling inside the collector tube [11]. With permission

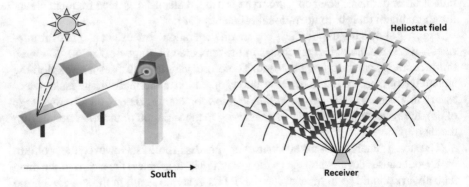

Fig. 4 Optimization steps of a Central Receiver System. Flux distribution prediction (left) and heliostat field optimization (right) [29]. With permission

The first one is the flat plate collector, shown in Fig. 5. Since this device requires no specific components, it is easy to manufacture them even at home. Students often build them from empty aluminum cans [47, 48] with a nickname of beer collector. The glass cover produces greenhouse effect to enhance solar flux utilization. These units

(a)

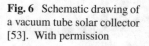

(b)

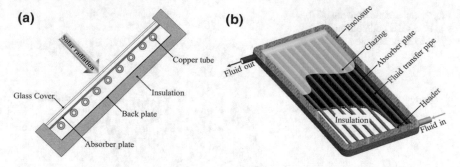

Fig. 5   Schematic drawing of a flat plate solar collector [53]. With permission

Fig. 6   Schematic drawing of
a vacuum tube solar collector
[53]. With permission

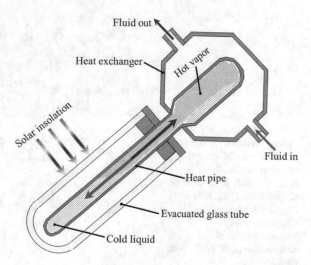

usually have low performance [49, 50], compared to other solar collector designs. However, this technology is excellent for drying food or seeds [51, 52]. It is worth to mention an old and simple variant which is used in weekend houses where there is no warm water. A large barrel is painted black, and the Sun heats the water up inside that is enough for a short shower for few people at the end of the day.

The vacuum tube solar collector, shown in Fig. 6, is an advanced domestic solar collector variant which features few expensive technologies. They are the evacuated tube, the heat pipe, and the absorber material around the heat pipe. If the glass breaks accidentally, usually only the heat pipe can be reused which is not environmental friendly. In addition, the payback period significantly extends if a replacement is required. To enhance the solar flux utilization, they may come with a parabolic mirror plate. The optimal tilt angle setup of vacuum tube solar collectors is discussed by Tang et al. [54]. They offer higher operational flexibility and efficiency than flat plate collectors as they are less affected by the incidence angle of the Sun [55, 56].

Fig. 7 Schematic drawing of
a hybrid solar cell/collector
[53]. With permission

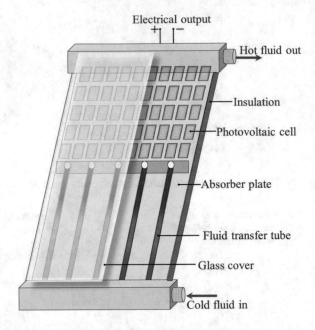

Fig. 8 Modeled efficiency
of vacuum (solid line) and
air-filled (dashed line)
Parabolic Trough Collectors
and their comparison with
experimental data. The
abscissa shows the inner
minus the ambient
temperature in °C [29]. With
permission

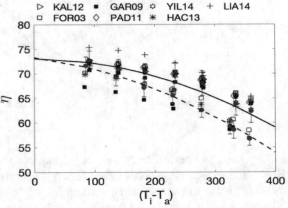

The last example is a hybrid solution where both electricity and heat is produced,
shown in Fig. 7. Currently, this variant is not competitive in price for domestic use,
however, it has great potential if the design can be efficiently integrated into buildings [57]. Since the efficiency of photovoltaic cells decreases with the increasing
temperature, this solution offers the advantage of cell cooling while the heat is also
used [58]. In addition to the flat plate design, the concept can be realized by using a
parabolic concentrator as well [59].

All solar collectors work most efficiently when their core temperature is close to
the ambient temperature, shown in Fig. 8 through the example of a PTC. This ensures
the least emitted heat to the ambient.

A slight increase in thermal efficiency of solar collector systems can be achieved by enhancing the heat transfer coefficient between the collector and the heat transfer fluid. It can be done by adding nanoparticles into the fluid [60, 61] which is called nanofluid. Even though the physical effect is present [62], it is still unclear that this technology is commercially viable or not. Beyond its production cost, the poor long-term stability of these fluids [63, 64] and the increase in the pumping energy demand due to the enhanced viscosity [65, 66] are also drawbacks. Bubbico et al. [67] concluded that stainless steel is resistant to the corrosion caused by the nanoparticles which is an indirect effect on cost increase of such solar collector systems.

2.2 Solar Cells

By covering only 0.1% of the Earth's surface with solar cells with only 10% efficiency, all of our energy demand would be covered [68]. There are numerous research papers dealing with solar cell price reduction [15, 69, 70], however, its end-of-life treatment is not solved at the industrial level [71]. Nevertheless, there are promising technologies on the horizon [72–75].

The development of solar cells is focusing around three main fields. They are the efficiency increase [76–78], surface coating processes [79–81], and layer stability for long-term use [82–84]. The next boom in the solar industry will be the upscaling of perovskite solar cell production [85] which has been started intensively about ten years ago, and the first production facilities have been built. The advantage of this mineral is that the production requires around 100 °C instead of 400–1400 °C like in the case of silicon-based cells [86]. Even though the silicon cells have higher efficiency, the manufacturing cost of the perovskite cells is a fraction of that which is unbeatable [86]. Multijunction cells provide the outstanding efficiency [87, 88], approaching 50%. However, they are too expensive for large scale use, hence, efficient solar spectrum utilization with a single cell is a reasonable way of efficiency increase [89]. Figure 9 shows the solar absorption of a silicon cell, highlighting the physical limitations. The structure and absorption range of a triple-junction solar cell is also shown.

From thermal engineering point of view, proper thermal control of solar cells can be done to ensure high efficiency [90], otherwise, there is not so much thermal aspect of this technology. As it was mentioned in Sect. 2.1, a hybrid solar cell/collector is an excellent choice for this purpose. The thermal balance of a solar cell is the following under steady-state operation:

$$\dot{Q}_{abs} = \dot{Q}_{in} - \dot{Q}_{refl} = P_{el} + \dot{Q}_{heat}, \tag{1}$$

where subscripts in order mean absorbed, incoming, reflected, and electric (power). The solar flux which is converted into heat heats the cell up while the ambient absorbs the emitted heat.

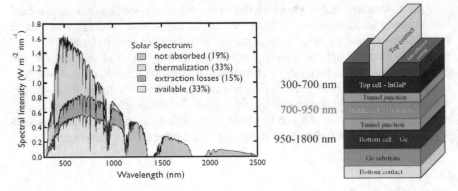

Fig. 9 Solar energy utilization by a silicon cell (left), and a triple-junction cell, indicating the wavelength absorption ranges of the layers (right) [89]. With permission

2.3   Example: Flat Plate Solar Collector

This subsection is divided into two parts. Firstly, a routine calculation is presented to determine the necessary solar collector area for domestic hot water production. Then the thermal balance of a flat plate solar collector is calculated, using a single node approximation which is the simplest method in thermal analysis.

The first step of solar collector sizing is the estimation of the daily hot water demand. If this is a family house or a smaller one, it is advised to have another source of hot water for appropriate system flexibility [91–93]. If the hot water demand is given, multiply it with a safety factor, i.e., by 1.2. Assume that the result is 500 l since it is a twin house for two families. This volume can be the same as the volume of the storage tank, V_t. Multiply the result with the specific heat of the water and the temperature difference between the hot and cold water that is usually around 40 °C to have high enough water temperature and low thermal losses. The heat stored in the tank is

$$Q_t = V_t \cdot \rho_w \cdot c_w \cdot \Delta T = 84 \text{ MJ} = 23.3 \text{ kWh}, \tag{2}$$

where $\rho_w = 1000$ kg/m^3 is the density, and $c_w = 4200$ J/(kg·K) is the specific heat of water. $\Delta T = 40$ °C was used. The next step is the search for the average global irradiation in online map databases. Take the three worst consequent months and calculate the average of them. For a slightly cloudy central European city, it is around $S_R = 2$ kWh/(m^2·day). The collector efficiency is always given as a chart by the manufacturer. Now let's assume that it is $\eta_c = 0.65$. The system efficiency is estimated as $\eta_s = 0.85$ that incorporates all the thermal losses in the pipe network of the building. Then the collector yield

$$C_y = S_R \cdot \eta_c \cdot \eta_s = 1.105 \text{ kWh/} \left(\text{m}^2 \cdot \text{day} \right). \tag{3}$$

Lastly, the required collector area can be calculated

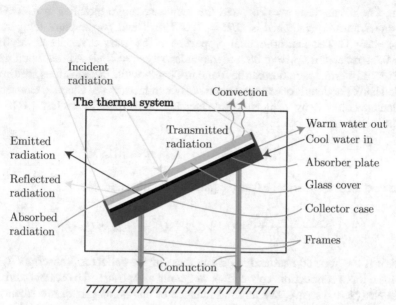

Fig. 10 The solar collector and its thermal connections to the ambient

$$A_C = Q_t/C_y = 21.12 \text{ m}^2. \tag{4}$$

This value means 100% solar energy coverage of the hot water demand. If it is reduced to, e.g., 60%, which is often suggested by the vendors, the resulting collector area will be 12.67 m². Finally, the best tilt and orientation of the collector is chosen, considering the orientation of the building and the roof structure. In the case of vacuum tube solar collector system, it has a relatively low sensitivity to the tilt angle unless it is orientated to south in the northern hemisphere or to north in the southern hemisphere.

The flat plate solar collector to be analyzed as a concentrated parameter model is shown qualitatively in Fig. 10. In this example, the analysis procedure will be detailed step-by-step, unlike in the case of later examples. The goal of this analysis is the determination of the collector efficiency and the mass flow rate of the produced warm water which can be stored.

The first step is sketching the problem, including the relevant bodies. Then the boundary of the thermal system is chosen. In this case, the system does not move or deform. Therefore, the boundaries can be fixed during the analysis. Presently, steady-state operation is assumed as all the boundary conditions are continuously present, and the daily variation is not important in this calculation phase. Hence, the mass and specific heat of the collector can be neglected; the only required material parameter for solid bodies is their thermal conductivity.

The collector in this example is assumed as 1 m wide and 2 m long, therefore, the absorbing surface area is $A_s = 2$ m². The thickness of the case is chosen as

0.1 m. The tilt angle is $\gamma = 30°$, and the incidence angle modifier is $K_\gamma = 0.95$. The direct normal irradiance is $DNI = 700$ W/m^2; and the ambient temperature is $T_\infty = 25$ °C. The thermo-optical properties of the glass cover are absorptivity, $\alpha_r = 0.04$, reflectivity, $\rho_r = 0.08$, and transmissivity, $\tau_r = 0.88$. A brief fundamental overview on thermal radiation can be found in Chap. 4 while the equations are directly applied here. For details on the fundamentals of conduction, see Chap. 2. Convection is calculated directly by using Newton's law. For fundamentals, see Ref. [94].

The incoming heat is

$$\dot{Q}_{in} = DNI \cdot A \cdot \cos(\gamma) \cdot K_\gamma \cdot \tau_r = 1014 \text{ W.} \tag{5}$$

In the next step, the thermal balance equation is written as

$$\frac{\mathrm{d}(c \cdot m_c \cdot T)}{\mathrm{d}t} = \dot{Q}_\lambda + \dot{Q}_h + \dot{Q}_r + \dot{Q}_{abs} - \dot{Q}_{in}, \tag{6}$$

where c is the specific heat and m_c is the mass of the collector. Subscripts λ, h, r, and abs refer to conduction, convection, radiation, and absorbed by water (sunk heat in the analyzed system). The thermal balance of any solid part of a concentrated parameter system should contain all these terms. Nevertheless, Eq. (6) can be—and usually is—simplified. The left side of the equation is zero since the present case is steady. The thermal conduction loss can be calculated by using the Fourier's law

$$\dot{Q}_\lambda = -\lambda_{al} \cdot A_{frame} \cdot \mathrm{grad}\,(T) \simeq \lambda_{al} \cdot A_{frame} \cdot \frac{T_{bp} - T_\infty}{\delta_{frame}}, \tag{7}$$

where δ_{frame} is the length of the frame and subscript bp notes the collector backplate. Assuming that the collector is installed on a roof with four 0.2 m long aluminum frames ($\lambda_{al} = 237$ W/(m·K)) with 0.001 m^2 cross-section each, and the backplate of the collector is $T_{bp} = 27$ °C, the resulting heat loss via conduction is $\dot{Q}_\lambda = 9.48$ W. Since this value is <1% of $\dot{Q}_{in}$, such terms are usually neglected. Nevertheless, the present calculation will feature it. By using warm water outlet temperature, $T_{w,o} = 65$ °C as a design parameter, the absorber surface temperature

$$T_a = \frac{\dot{Q}_{abs} \cdot \delta_a}{\lambda_{al} \cdot A_s} + T_{w,o} = 65.1 \text{ °C,} \tag{8}$$

where $\delta_a = 0.1$ m was assumed as an effective conduction length. Since $\dot{Q}_{abs}$ is a result that depends on later calculation results, iteration is required. For an initial guess, $\eta_c = 0.73 = \dot{Q}_{in}/\dot{Q}_{abs}$ was used from Fig. 8. Due to the high conductivity and low characteristic length, a single iteration step was enough to get T_a. In reality, T_a has a spatial distribution, and thermal conduction between the absorber surface and the glass heats this latter part up since there is air between them. This complex phenomenon is now omitted, and to compensate the conduction loss, it is assumed that $T_a \sim T_{w,o}$. Next, thermal radiation is calculated which begins with the determination of the temperature of the glass, T_g.

$$Q_{r,a} = \sigma_0 \cdot \varepsilon_a \cdot \varphi_{a \to g} \cdot A_s \cdot \left(T_a^4 - T_g^4\right) = \sigma_0 \cdot \varepsilon_g \cdot \varphi_{g \to \infty} \cdot A_s \cdot \left(T_g^4 - T_\infty^4\right), \quad (9)$$

where $\sigma_0 = 5.67 \cdot 10^{-8}$ W/(m^2·K^4) is the Stefan-Boltzmann constant, the view factors, $\varphi_{a \to g}$ and $\varphi_{g \to \infty}$ are assumed as unity, and the emissivity of the absorber, ε_a, is also unity while that of the glass is 0.9 in the infrared regime where these bodies emit the most heat [95]. Solving for T_g gives 48 °C. Then the radiation loss of the absorber plate can be calculated which is 278.3 W. Finally, convection losses are determined as

$$\dot{Q}_h = h \cdot A_s \cdot (T_h - T_\infty), \quad (10)$$

where T_h is the effective temperature of the collector from thermal convection point of view. The surface area of the case is 2.6 m^2, hence by multiplying T_{bp} with this area and T_g by A_s, then dividing by the total surface area, $T_h = 36$ °C is the result. Note that this simple weighting could be done because Q_h linearly scales with T. The heat transfer coefficient, h, is calculated by using Eq. (5) from Ref. [96], Subsection F2.2:

$$h = \left[0.56 \cdot (\text{Ra}_c \cdot \cos{(\gamma)})^{1/4} + 0.13 \cdot \left(\text{Ra}^{1/3} - \text{Ra}_c^{1/3}\right)\right] \cdot \lambda_{air}/L = 3.25 \text{ W}/\left(\text{m}^2 \cdot \text{K}\right), \quad (11)$$

where $\text{Ra}_c = 10^8$ is the critical Rayleigh number at $\gamma = 30°$ [96], $\lambda_{air} = 0.02624$ W/(m·K) is the thermal conductivity of air, and $L = 2$ m is the characteristic length of the solar collector. Ra is the Rayleigh number, calculated by Eq. (12)

$$\text{Ra} = g \cdot \beta_{air} (T_h - T_\infty) \cdot L^3 / (\nu_{air} \cdot \alpha_{air}) = 7.5 \cdot 10^9, \quad (12)$$

where $g = 9.81$ m/s^2 is the gravitational acceleration, $\beta_{air} = 0.00338$ 1/°C is the thermal expansion coefficient of air, $\nu_{air} = 15.52$ mm^2/s is the air kinematic viscosity, and $\alpha_{air} = 21.95$ mm^2/s is the air thermal diffusivity. Hence, $\dot{Q}_h = 166.2$ W is the conductive loss. A final term, the radiation loss of the backplate is calculated

$$Q_{r,bp} = \sigma_0 \cdot \varepsilon_{bp} \cdot \varphi_{bp \to \infty} \cdot A_{bp} \cdot \left(T_{bp}^4 - T_\infty^4\right) = 15.8 \text{ W}, \quad (13)$$

where $\varepsilon_{bp} \cdot \varphi_{bp \to \infty} = 1$. By subtracting all the losses from the incoming heat power the absorbed heat by the water is

$$Q_{abs} = \dot{Q}_{in} - \dot{Q}_{r,a} - \dot{Q}_{r,bp} - \dot{Q}_h - \dot{Q}_\lambda = 543.9 \text{ W}. \quad (14)$$

Hence, the collector efficiency is $\eta_c = 0.54$. The mass flow of water that can be continuously heated up to 65 °C under these conditions is

$$m_w = \frac{\dot{Q}_{abs}}{c_w \cdot \left(T_{w,o} - T_\infty\right)} = 3.24 \text{ g/s}, \quad (15)$$

where $c_w = 4200$ J/(kg·K) is the specific heat of water. Real pumps have low efficiency in such a small size, therefore, larger ones are used which operate intermittently. In addition, the heat-storing procedure is also depending on the layout of the collectors. Note that Eq. (11) was calculated by assuming still conditions. However, when the wind starts blowing, the heat transfer coefficient significantly increases. At 1 m/s $h = 5$ W/ $\left(\text{m}^2 \cdot \text{K}\right)$ while $h = 13.2$ W/ $\left(\text{m}^2 \cdot \text{K}\right)$ at 3.5 m/s. Hence, $\dot{Q}_h$ in Eq. (6) becomes too large, and the temperature of the water will not reach the desired $T_{w,o} = 65$ °C. Since this wind speed is quite common in central Europe, solar collectors are not effective as in this region as in southern landscapes, and the return period of the investment becomes excessive compared to other renewable technologies. Nevertheless, the latter wind speed is high enough to start spinning up a wind turbine which is discussed next.

3   Wind Turbines

Wind turbines are the most iconic representatives of the revolution in the energy industry towards a sustainable future. Beyond numbers on their efficiency and generated power, its social acceptance is a critical part before deploying a unit/farm [97–101]. To ensure the trust of the public in the technology, the possibility of injuries and damages in civil buildings and infrastructure must be assessed and considered during the selection of location [102–104]. The most notable impact of wind turbines on the people living nearby is the generated noise by the blades which may lead to discomfort and health issues [105–108].

When the society supports the building of wind turbines, its best location has to be selected, considering multiple factors [109–111]. This is the phase where the decision can be made based on objective numbers. The first step is considering the wind potential of the location [112–114]. Besides the cost of the product, the cost of foundation [115–117] and installation [118, 119] is also considerable. In addition, transportation may be challenging as well [120]. Another aspect of wind turbine installation is its impact on biodiversity that should also be taken into account [121–123].

Wind energy is emerging as a renewable power source in the mainlands since the relatively low price of fossil fuels favors thermal power plants. Nevertheless, the increasing carbon dioxide allowances [124] slowly turn the policies. The situation is intensified in islands where the high cost of fossil fuel transportation makes the use of renewable energy an economically preferred option. Hence, these regions are the incubators of sustainable technologies [121, 125–127]. Figure 11 shows a wind farm in Faial Island, a member of the nine islands of the Azores.

Wind turbines are usually installed in farms, consisting of numerous wind turbines. Even though the wind potential is known, designing the optimal layout is not a trivial task as every turbine has a wake that reduces the extractable power by all the turbines downstream [128–130]. Using advanced algorithms, it is a possible outcome of an

Fig. 11 Wind farm in the Azores

analysis that the maximum wind energy utilization can be achieved by shutting down some turbines in an existing wind farm under certain circumstances [131].

The cause of relatively low wind speed above the ground is the atmospheric boundary layer which is also disturbed by natural and artificial obstacles, leading to lower power generation [132–135]. A solution to this problem is building flying power plants as the wind speeds are significantly higher above. Even though there are several excellent ideas present [136], none of them was able to achieve an international breakthrough yet; the market is dominated by the horizontal axis wind turbines.

The generated power by wind turbines can be calculated as

$$P = C_P \cdot \frac{1}{2} \cdot \rho \cdot v^3 \cdot A, \tag{16}$$

where P is the output (electric) power, C_P is the power coefficient, ρ is the density of air, v is the wind speed, and A is the area of the rotor. According to Betz's law, the maximum value of C_P is 16/27 which cannot be increased by any means. It is also an outcome of Betz's law that the wind cannot be stopped. In wind turbine literature, all the losses are put into C_P, hence, the reader rarely can find detailed information on the efficiencies of various parts. From the theoretical point of view, C_P is the overall efficiency of the unit, considering both theoretical and practical limitations.

3.1 Thermal and Mechanical Diagnostics of Wind Turbines

Equation (16) tells that if there are ideal components used in the wind turbine, energy generation can be realized as an isothermal process. Hence, there would be no reason to perform thermal analyses. This finding anticipates that thermal-related problems do not necessarily originate from design issues. Instead, they are the cause of wearing components which approach their end of life; this is the motivation of this subsection. Consequently, excessive temperature at any moving part of the wind turbine apparently originates from a mechanical or electrical defect. Since no personnel is required to operate a wind turbine, diagnostics is critical to facilitate problem mitigation before spectacular unit failures. Today, it means online monitoring systems to allow the use of each component until they approach their end of life. To ensure reliable operation when they are aged [137], diagnostics and control algorithms are connected [138–140]. A spectacular example for this phenomenon is the effect of dust, ice, or insect accumulation on the blades [141, 142] which notably decrease the output power, even if a thin layer is covering the blades.

Figure 12 shows the main components of a horizontal axis wind turbine which will be analyzed from thermal point of view in Sect. 3.2. The pitch angle of the blades is the most important parameter which allows the control of the extracted power from the wind [143]. It is set to a very steep angle, where the blade chord is almost parallel to the tangential velocity, to utilize the maximum power until the design point is reached [144]. Then the pitch angle is increased to decrease the load of the system in high-speed winds, maintaining the nominal electric output power. There is a limiting wind speed above which the pitch angle is set to 90° to survive storms [102]. To increase the operational flexibility of pitch angle control, fault-tolerant algorithms can be used [145, 146].

All the further noted parts in Fig. 12 along the axis have losses which finally turns into heat that increases the temperature of the components. The efficiency of bearings typically exceed 99% [147], overloading, cracks, and material failure all causes deformations which lead to wear. Hence, its efficiency decreases, and operating temperature increases [148]. At elevated temperature, the lubricating grease might flow out, and the wearing process speeds up, shortly resulting in shaft stopping if not detected and counteracted in time. For its lower failure tendency, recent wind turbines feature tapered roller bearings [149]. The cheapest way to detect bearing failure is the installation of a temperature sensor on its housing as all damages lead to increased heat generation in this part. It even helps in detecting misaligned shafts [150].

The most complex part of the wind turbine is the gearbox. They typically contain planetary gear sets with ~1.5% loss per stage [151]. Its full-scale measurement, including thermal and mechanical aspects, was discussed by Fernandes et al. [152], analyzing three power levels. Its condition monitoring is critical to find material and mechanical defects [153–155]. Even though the efficiency of the gearbox is high, it has a relatively small surface area, hence, it may require liquid cooling besides convection by air. The high temperature principally lowers the oil viscosity which may then provide insufficient lubrication, leading to increased wear of the contacting surfaces.

The electric generators also feature bearings which might cause operational problems. Beyond mechanical issues, electromagnetic vibration can also lead to failures [156]. The principal heat source of the generators is the Joule heating of the electric wires, which has to be considered during the design procedure. To mitigate this effect, superconductors can be used, however, there are still several issues to solve before its application in a wind turbine [157]. The degradation of insulation is a notable

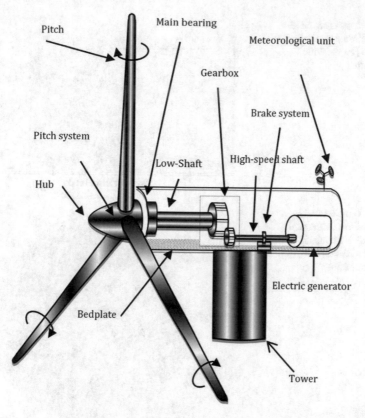

Fig. 12 Main mechanical components of a horizontal axis wind turbine [143]. With permission

source of generator failures. Nevertheless, online frequency response analysis with appropriate filtering is a suitable tool for early detection [158].

In the case of every complex and expensive machine, the operation should include preventive mechanisms to avoid excessive costs of repairing. However, there is a cost optimum between fully preventive and repair-based maintenance strategies, shown in Fig. 13.

To address the tendency of each component to failure, there are three approaches available. They are failure analysis, reliability analysis, and risk analysis [159]. If there is enough data, Fuzzy logic-based analysis also can be performed [160] to highlight design problems. In addition to focusing on past failures, the estimation of the fatigue load is also an excellent approach [161] to evaluate operating turbines.

An overview of common failures is shown in Fig. 14. Since the energy conversion of a wind turbine can be assumed as an isothermal process, there is no direct reference to thermal issues. If this is approached from the other side, i.e., which problem might not originate from thermal issues, only the yaw system and rotor blade failure categories can be named. Therefore, thermometers, as they are the cheapest sensors,

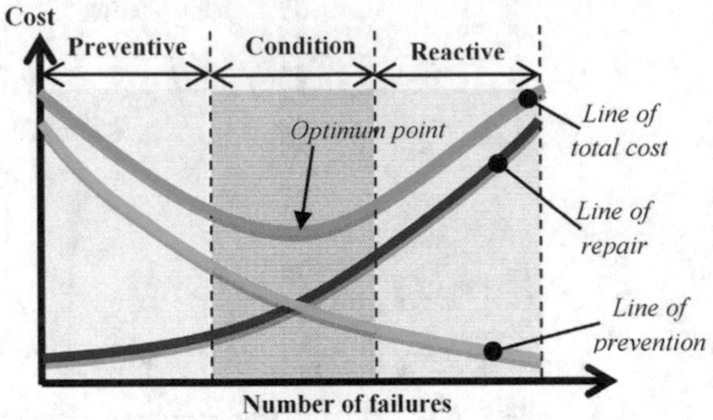

Fig. 13 Costs associated with traditional maintenance strategies [143]. With permission

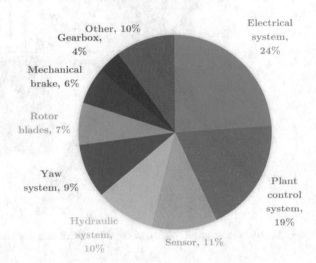

Fig. 14 Typical wind turbine failure causes and their relative frequency. Data reproduced from [143]. With permission

are suggested to put on several locations across the turbine to allow the monitoring of each component. Redundancy is always advised for critical parts.

3.2 Example: Thermal Network Analysis of a Wind Turbine

A 136 kW horizontal axis wind turbine is analyzed where the nominal power is reached at 8 m/s wind speed. The aerodynamic losses are now neglected since this example focuses on thermal issues. Figure 15 shows the Sankey diagram of the analyzed system and Table 1 summarizes the efficiencies of the highlighted components, including the limitation by Betz's law. The resulting C_P is 0.489, which is a high value and characterizes modern wind turbines [162, 163].

This example will be solved by using the thermal network (TN) modeling approach. Recalling Chap. 1, it is a concentrated parameter method where the system is divided into several subcomponents which are handled as a single node. The heat

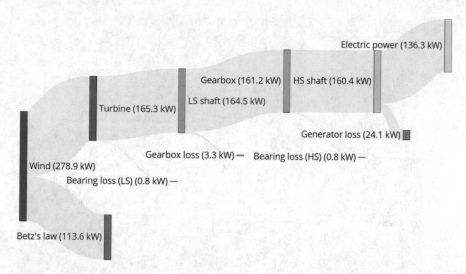

Fig. 15 Sankey diagram of the example wind turbine. HS and LS mean high-speed and low-speed, respectively

Table 1 Efficiencies of the wind turbine components

Betz limitation	16/27
Thrust bearing	0.995
Gearbox	0.98
Generator	0.85

transfer is evaluated between them and the ambient. Here, the system was divided into five parts, and heat generation is present in four of them. They were the front part of the nacelle (FN), the gearbox (GB), the rear part of the nacelle (RN), and the generator (GE). Since thermal radiation plays a notable role, the cover (CO) was the fifth part. The calculations were performed in Matlab Simulink software environment which also allows a spectacular representation of the problem to be solved.

Figure 16 shows an overview of the system. There are three columns with several subparts which were colored to provide a smoother workflow. For simplicity, all the variables were linked to the appropriate boxes by using a from-goto pairs. Hence, if a variable has multiple occurrences, this solution allows a wire-free work, resulting in a cleaner workspace. The top left box contains the input efficiencies of Table 1 and the input power. The output parameters of this box are the volumetric heat generations. The input power is analyzed as a constant value and a temporal function, using real wind speed data. It should be highlighted at this point that Eq. (16) contains the cube of the wind speed. Therefore, rough and linear estimations both fail in modeling the wind turbine output power when sparse data is available. There is no internationally accepted guide for the sampling frequency, however, 10–15 s might be enough for a fair estimation. Consequently, free online weather databases, which contain 3 h

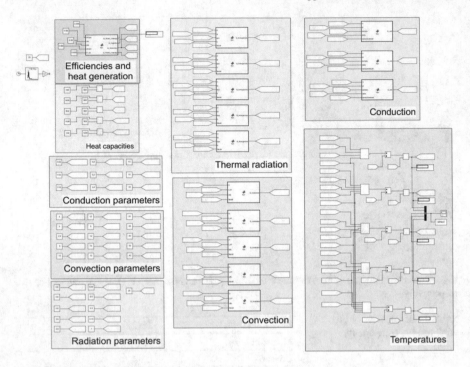

Fig. 16 Overall view of the thermal network model

data, are useless for mapping the exact wind potential and evaluating the economic feasibility of building a single wind turbine or a farm.

The yellow area summarizes the heat capacities, i.e., the product of the mass, m_c, and specific heat, c, of various components. It is assumed that they do not change over time, hence, Eq. (17) has to be solved for each node:

$$\frac{\mathrm{d}T}{\mathrm{d}t} = \frac{1}{c \cdot m_c} \cdot \left(\dot{Q}_v + \dot{Q}_\lambda + \dot{Q}_h + \dot{Q}_r \right), \tag{17}$$

where $\dot{Q}$ is the thermal power. Subscripts v, λ, h, and r refer to volumetric heat generation, conduction, convection, and radiation, respectively. Light blue, red, and green boxes represent thermal conduction, convection, and radiation. The parameters were defined in the first column, and the calculation of the resulting thermal powers was performed in the second and third columns which have the same color. The heat conduction equation is

$$\dot{Q}_{\lambda,1 \to 2} = -\lambda \cdot A_\lambda \cdot \frac{T_2 - T_1}{\delta} = \lambda \cdot A_\lambda \cdot \frac{T_1 - T_2}{\delta}, \tag{18}$$

where λ is the thermal conductivity, A_λ is the effective contact area of the two bodies with T_1 and T_2 temperatures. The last term is the temperature gradient over a δ distance between the bodies. Convection is calculated as

$$\dot{Q}_{h,1\to\infty} = h \cdot A_h \cdot (T_1 - T_\infty), \tag{19}$$

where h is the heat transfer coefficient; ∞ denotes the far-field temperature near part 1, and A_h is the effective area of convection. Following Ohm's law of electric conductivity, a similar equation can be formed for conduction and convection, i.e.,

$$\dot{Q} = \Delta T / \theta, \tag{20}$$

where θ is the thermal resistance. It is $\theta_\lambda = \delta / (\lambda \cdot A)$ and $\theta_h = 1 / (h \cdot A)$ for conduction and convection, respectively. Since both processes proportional to T, this parameter might simplify the problem-solving. However, it is omitted in the present analysis since radiation is also significant where $\theta_r = \theta_r (T^3)$, which does not help a lot and might easily lead to model building user mistakes when the system is large. Nevertheless, the use of thermal resistances is reasonable when there is contact resistance present between two surfaces. In this case, the conductivity of the thermal resistance, similar to the dimension of h, is introduced. The contact resistances are omitted in the present analysis. If such a problem occurs, there are various thermal greases available on the market. There are products which are also good electrical conductors and others which are good electrical insulators. Thermal radiation is calculated as

$$\dot{Q}_{r,1\to2} = \sigma_0 \cdot \varepsilon_{1\to2} \cdot \varphi_{1\to2} \cdot A_1 \cdot (T_1^4 - T_2^4), \tag{21}$$

where ε is the common emissivity, and φ is the view factor. For more details on thermal radiation see Chap. 4. Tables 2, 3 and 4 summarize all the used parameters in this calculation.

Since the model is complete, thermal analysis can start. 20 °C ambient temperature is assumed in all the presented cases. Firstly, the steady loading is analyzed at 10% and 100% of the rated power. Figure 17 shows the first case. Here, the maximum temperature is 65 °C, which is reached by the generator since it has the lowest efficiency. Hence, the volumetric heat generation is the highest at 1.765 kW. As for the other parts, the generated heat is practically absorbed by the ambient via natural

Table 2 Masses, heat capacities, and heat transfer parameters of the model

Part	m (kg)	c (J/(kg·K))	A_h (m^2)	h (W/(m^2·K))	$T_{\infty,h}$ (°C)
FN	140	500	3	12	20
GB	1000	500	3	5	25
RN	1000	500	3	12	35
GE	800	500	2.5	12	40

Table 3 Parameters of thermal conduction calculation

Contact	λ (W/(m·K))	δ (m)	A_λ (m^2)
FN→GB	15	0.2	0.03
GB→GE	15	0.2	0.03
GE→RN	15	0.2	0.15

Table 4 Parameters of thermal radiation. amb denotes the ambient

Contact	$\varepsilon_{1\rightarrow2}$	$\varphi_{1\rightarrow2}$
FN→CO	0.8	0.05
GB→CO	0.8	0.5
GE→CO	0.8	0.3
RN→CO	0.8	0.15
CO→amb	0.5	1

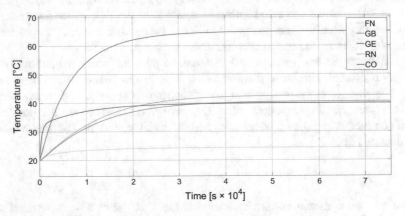

Fig. 17 Temperature history of all nodes at 10 kW electric power

cooling. There is an error in this calculation, namely, the temperature of CO that exceeds that of GE in the initial phase. This is due to the constant assumed temperature of the far-field in the calculation of convection. There are numerous ways to eliminate this non-physical behavior, e.g., by defining this temperature based on the generator temperature. However, this is omitted now in favor of simplicity.

Figure 18 shows the temperature distribution at a constant 100 kW electric power operation. The trends are similar, however, the maximum temperature here is 240 °C. This value is not high for metal parts, while plastics may melt at this temperature, including insulating materials, and if the generator has permanent magnets in the rotor, their temperature has to be checked to remain below the demagnetization temperature. Nevertheless, roughly 8 h are required to get to the final temperature. In most regions in the mainland, this constant wind loading is rare, hence, the wind turbine is analyzed next with real wind temporal data. Nevertheless, a wind turbine has to survive the long presence of its nominal loading.

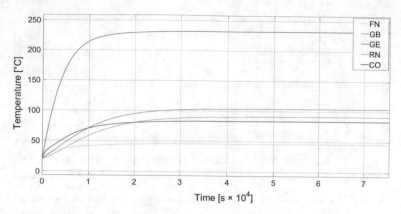

Fig. 18 Temperature history of all nodes at 100 kW electric power

Fig. 19 Sample weather data at LHKV airfield, downloaded from the daily database of idokep.hu. Range: 15 December 13:00–16 December 10:00, 2018. Data was made available by Cirrus Hungary Kft.

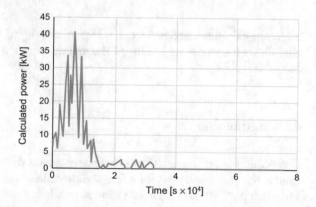

The used wind data was converted into electric power, using $C_P = 0.489$ and rotor diameter of 34 m. The result is shown in Fig. 19. The amplitude of wind speed fluctuation was low, however, $P \sim v^3$ amplified it. The power was zeroed below 2 m/s. According to the data, there was a 5 h period where the wind was strong enough to drive the turbine; after that, it was practically zero for the rest of the day.

The 5 h operation at various power proposes that the maximum generator temperature will remain well below 240 °C. The results are shown in Fig. 20. The peak temperature here is 120 °C which then approaches the T_∞ temperatures, shown in Table 2.

It can be concluded that the model worked appropriately and provided the desired temperature maps. However, in reality, the heat is generated in small parts which are in contact with larger parts. Hence, further elements can be introduced to the calculations to get a more realistic temperature distribution of the critical parts. Such a model is excellent to perform preliminary analyses based on the expected location of a single wind turbine or a wind farm since the run time on a single processor thread was about 1 s. In addition, the operating parameters can be adjusted based

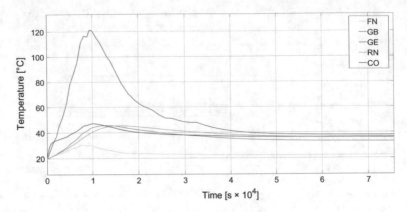

Fig. 20 Temperature history of all nodes at real wind conditions

on sensor data, discussed in Sect. 3.1. Hence, this digital twin of a real wind turbine also allows the proper scheduling of the maintenance if a component approaches its end of life.

4   Combustion

Combustion is far the most complex phenomenon discussed in the present book. Hence, this paragraph contains a few comprehensive references for more details on various topics. Besides thermodynamics and heat transfer, it involves principally chemistry [164–166]. Due to the practical scales, combustion can be modeled [167] by using the continuum-based physical approach [168, 169]. To understand the phenomenon from the application side, see Refs. [170–172]. Since gas turbines are the most complex heat engines produced in a large volume today, understanding their principles and challenges helps a lot in other applications as well [173]. For a detailed definition of combustion-related terms and properties, see Ref. [170].

4.1   Why Combustion Is Still Important?

Fossil fuels accounted for roughly 86% of the primary energy use in 2015 [174] with increasing utilization tendency. However, the will is present to substitute fossil fuels with renewable ones [175], there are no economically viable alternatives present, even though fuel cell and electric technologies are rapidly developing [176, 177]. This trend, along with the introduction of other renewable power technologies, is pushed by the rapidly increasing CO_2 allowance price [124]. The high energy density in both mass and volume is crucial in aviation. Hence, the long-term strategy seems to be the use of various biofuels [178, 179]. Nevertheless, this scenario raises the question of food security [180]. Proper alteration of the oxygen content of the air practically eliminates NO_X emission. A notable technology for this purpose is oxy-fuel combustion where the combustion temperature can be high for increased thermal efficiency [181, 182], and furthermore, the CO_2 content of the flue gas can be easily separated and stored instead of emitting it to the atmosphere. Another solution is the recirculation of the flue gas to dilute combustion air, leading to notably lower overall oxygen concentration in the oxidizer medium side while its temperature is increased, facilitating the ignition of the fuel. There are several names available in the literature for this concept; the most widespread one today is flameless combustion [183]. To avoid carbon dioxide emission, ammonia combustion gained increased attention recently [184, 185].

As a consequence, it is certain that combustion engines will remain in service for long decades. In addition to technical challenges, it also takes a long time to build the necessary infrastructure and support services for electric city cars only; the situation in long-distance transport is a more complex problem [186] as current battery technologies provide roughly one hundredth energy density of that of hydrocarbon fuels. Consequently, the authors found it important to discuss combustion in the framework of the present book as well.

4.2 Combustion Aerodynamics

Before jumping into the details of the combustion process, the present subsection briefly summarizes the corresponding aerodynamics [187]. Its importance was first recognized by the Sorbonne University who asked Theodore von Kármán to give a lecture on combustion for the 1950/51 school year. Prior that, this science consisted of the chemistry of flames on one side, and applied engineering through trial and error on the other side. Aerodynamics was the bridge between the two fields, hence, giving birth to modern combustion science. Even though the computational tools [188–190] and the laser measurement techniques [191–194] show an ongoing rapid development, the foundation of principles and approaches date back to the '50s.

Steady combustion needs a continuous supply of fuel and oxidizer; while the location of the flame front depends on the aerodynamics. If the flame propagation velocity is exceeded by the velocity of the combustible mixture, the flame is pushed downstream. When there no equilibrium position is reached, the flame will be blown off. The other case, when the flame propagation velocity is larger, the flame front moves upstream, called flashback. Either of the two phenomena may cause severe damage to the equipment.

Figure 21 shows a section of a simple, modern burner. Fuel and air are mixed before the inlet to minimize the pollutant emissions through homogeneous combustion. A set of stator blades, called swirl vanes, add tangential velocity component to the fuel-air mixture to form the flow field downstream. The central flame stabilizer geometry is responsible for providing the appropriate mixture velocity that exceeds

Fig. 21 Cross-section of a
gas-fired lean premixed
swirl-stabilized burner (left)
and a picture of a
liquid-fueled swirl burner
(right). The yellow flares are
caused by the burning
droplets

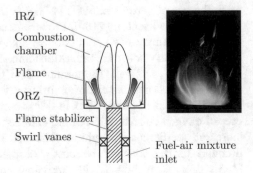

IRZ

Combustion
chamber

Flame

ORZ

Flame stabilizer

Swirl vanes

Fuel-air mixture
inlet

the flame propagation velocity to prevent flashback. The combustion chamber features an increase in the cross section to allow ignition and provide enough room for a stable flame. The presented flame stabilizer is a bluff body, which means that there is a flow separation in its wake. It is the Inner Recirculation Zone (IRZ) in Fig. 21. The conservation of momentum leads to the formation of a counterpart, the Outer Recirculation Zone (ORZ). Their role is to recirculate hot flue gas and a small concentration of radicals which facilitate the ignition of the cold, fresh mixture. When liquid fuel is utilized, it has to be atomized first [195–197] to facilitate the evaporation of droplets [198] and mix well the fuel vapor with air to achieve homogeneous and clean combustion.

The aerodynamics of the burner determines the flame shape, hence, the local concentration of the fuel and oxidizer, the heat release rate, and emissivity. Consequently, understanding the burner operation is necessary prior to performing any thermal calculation on combustion chambers.

4.3   Governing Processes of Combustion

The aim of this subsection to provide the necessary knowledge of reaction kinetics to thermal engineers to better understand this complex process. Instead of going into deep details of the chemistry, only key phenomena are discussed here which help in the understanding of the resulting thermal conditions. If there are large forest which contains several combustible species under the open atmosphere that is rich in oxygen, why a fire doesn't start immediately?

In reaction kinetics, there is a 1000 K limitation in temperature below which combustion is considered as low-temperature, and above that, it is high-temperature [199, 200]. The molecular structure plays a highlighted role in the lower regime, causing, e.g., the knocking of internal combustion engines [201]. This section will continue with high-temperature combustion since it is more common in the practice. In this regime, principally the atomic composition counts. Hence, the combustion characteristics of a fossil-originated fuel will be identical to that of a renewable fuel.

Therefore, only the mass fraction of C, H, O, N, and S atoms, ash, and water content of the fuel count in this case.

The fundamentals are presented via the net reaction of methane oxidation:

$$CH_4 + 2O_2 \longrightarrow CO_2 + 2H_2O + \Delta H, \tag{22}$$

where ΔH is the net heat of combustion. This quantity is available for various species in online databases and textbooks [202, 203]. If CH_4 is oxidized with air, the appropriate amount of N_2 can be added to both sides. Equation (22) is true if combustion is perfect, so all the fuel and oxidizer are consumed. However, net reaction equations are useful only when the combustion chamber is treated as a black box. Even though fuel and oxidizer are mixed, there are further limitations of combustion. These are:

- The concentration of the mixture is within the corresponding flammability limits, i.e., if the mixture contains either fuel or oxidizer in a low concentration, the reactions will stop and not initiate a chain reaction.
- There is high enough temperature to start the reactions. The ignition source can be a spark [204] or a hot surface/gas to provide the necessary activation energy.

Therefore, chemical reactions do not occur instantaneously. The reaction rate, r, controls the heat release as follows:

$$r = k\,(T)\,[A]^m\,[B]^n\,, \tag{23}$$

where $k\,(T)$ is the reaction rate constant with temperature dependence. $[A]$ and $[B]$ are the molar concentrations of the reactants. m and n are the partial orders of a reaction which are determined experimentally. Note that they differ from the stoichiometric coefficients. Reference [164] is recommended for further reading on this topic. k typically follows the Arrhenius equation:

$$k = A \cdot \exp\left(\frac{-E_a}{\Re \cdot T}\right), \tag{24}$$

where A is the pre-exponential factor which is often multiplied by a power of T. E_a is the activation energy and $\Re$ is the universal gas constant.

The real oxidation of CH_4 in air, considering pollutant formations, such as NO_X, is crucial in many practical applications as this is the core of all hydrocarbon mechanisms. The state-of-the-art reaction kinetic mechanism in this field contains 153 different species and 1639 reaction steps [205]. Hence, all the parameters of the above equations need to be determined for each reaction, meaning a rather complex system.

Anyone can download these reaction mechanisms or kindly ask the authors of a paper to provide it, however, they are rarely practical unless a complete mechanism is really necessary. To model combustion in, e.g., a computational fluid dynamics code, simplifications are necessary since the transport equations need to be solved for all species in all cells in each step which easily lead to excessive computational time. Here are few key tips to domesticate these large systems.

1. Know your mixture. If it is rich in fuel, even locally, the following reaction types should not be omitted, using CH_4 as an example.

$$CH_4 + M \longrightarrow CH_3^* + H^* + M, \tag{25}$$

$$2CH_3^* \longrightarrow C_2H_6, \tag{26}$$

where the asterisk denotes the radicals. Therefore, more complex species and their oxidation mechanism should be considered in rich mixtures than the initially available ones. This is the way how larger species form, finally leading to the formation of soot particles in the absence of oxygen. The exact formation of soot particles is under research, see Refs. [206–208] for further information on this topic. Due to the complex structure of soot, it is usually modeled separately from the main reactions.

If the mixture at the inlet port is lean enough, i.e., there is at least 3% more oxygen present than it is necessary for the stoichiometric reactions, the mixture can be considered as lean. In this case, the fuel species can be completely consumed and the available thermal energy is released. Consequently, the oxidation of the fuel species dominate the reactions and more complex species can be fully omitted during the simulation. For CH_4, its oxidation into CO_2 is the following.

$$CH_4 \longrightarrow CH_3^* \longrightarrow CH_2O \longrightarrow CHO^* \longrightarrow CO \longrightarrow CO_2 \tag{27}$$

It can be seen—which is quite general in hydrocarbon combustion—that the hydrogen will be released at first while the last step is the oxidation of CO into CO_2. Therefore, if the oxidizer concentration does not satisfy the conditions of complete combustion, the CO content will rise in the flue gas at first which is then followed by hydrocarbon emission in various forms, depending on the oxygen concentration in the combusting mixture. When a long-chained hydrocarbon is the fuel, highly toxic aromatic compounds are easily formed, and there will be blue a smoke visible downstream the flame.

The importance of the mixture properties is their direct effect on the radiation characteristics of the flame which is discussed in Sect. 4.4.

2. If air is used as an oxidizer, it is good to know that NO_X formation is a slow process with long time scales. If its formation is of secondary importance, the oxidation of N can be fully omitted. Further information on this process is available in the literature [209].

3. A reaction mechanism can be reduced to form lumped reaction steps. The principal aim is the reduction of the number of species [166]. Consequently, this procedure often accompanied by loss of information, leading to uncertainties. The goal needs to be clearly set to end up with the most suitable result. If the reduced number of species is too low, the uncertainty of the computational model will be

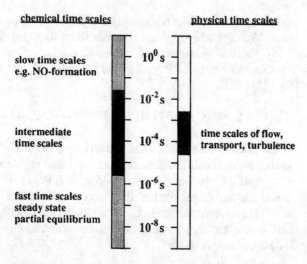

Fig. 22 Time scales of chemical and fluid dynamical processes [213]. With permission

high. The reduction procedure needs key parameters from the application side at first which should be followed more strictly. In thermal engineering, the most important parameter is the heating value. However, ignition delay time is also crucial for combustion simulation to properly estimate the shape of the flame. If a specie, e.g., OH* and/or CH* is measured by a spectroscope [210–212], these should be calculated with high accuracy. Hence, for a given computational power available, the result is a compromise between spatial resolution (mesh size) and mechanism uncertainty.

4. Typical hydrocarbon reactions in lean mixtures happen in a significantly shorter time scale than the flow time scale [213], shown in Fig. 22. To quantify this, Damköhler numbers are introduced which express the ratio of various timescales. From modeling point of view, the flow-to-chemical time scale ratio is the most important one, $Da = t_{flow}/t_{chem}$. Here, t_{flow} is the turbulent integral timescale, and t_{chem} is the time timescale of the set of chemical reactions of interest. If $Da > 1000$, the progress of the combustion from its initial components towards flue gas can be characterized by a single variable, named progress variable. This is automatically calculated by software [214] to streamline the workflow of engineers. Hence, it is advantageous from one side as the reaction kinetics is replaced by an automatic lookup table with a low number of species. Nevertheless, modifications on it, considering additional species, steps are hard to perform, and it is usually better to ask a reaction kinetics expert to provide the proper reduced mechanism.

4.4   Thermal Calculations

For any thermal calculation, the most important parameter of the fuel is its higher heating value, *HHV*. It is the available heat when the fuel is completely combusted, and all the products are cooled down to the initial temperature, and H_2O is in liquid form. This quantity does not consider dissociation due to high flame temperature [165]. There were numerous formulae developed for various fuels, nevertheless, the most versatile one which can be used for practical gaseous, liquid, and solid fuels is Eq. (28) [215]:

$$HHV = 0.3491C + 1.1783H + 0.1005S - 0.1034O - 0.0151N - 0.0211a, \quad (28)$$

where a denotes the ash content, and all components are understood in mass percentages of the dry fuel. The dimension of the result is [MJ/kg], and it is accurate within $\pm 3\%$, and the average error is $\pm 1.45\%$. See Ref. [215] for the detailed ranges on model applicability. Note that Eq. (28) can be used without calculating the number of bonds between the atoms. Lower heating value, LHV, is the quantity which is the heat of combustion when all products are cooled down to the initial temperature, and H_2O is present as steam

$$LHV = HHV - (9H + w) \cdot L_{H_2O}, \quad (29)$$

where H is the hydrogen, and w is the water content in the fuel. L_{H_2O} is the latent heat of vaporization of the water which is 2.51 MJ/kg under atmospheric pressure. In practice, the acid dew point limits the heat recovery from the flue gas, especially the condensation of sulfuric acid in coal-fired thermal power plants [216, 217]. Hence, it is sure that a certain portion of the heat is lost to avoid the excessive cost of additional treatment systems to be installed along with a corrosion-resistant chimney. The thermal efficiency, η_{th}, of a boiler or other combustion plant is

$$\eta_{th} = \sum Q_{abs} / \sum Q_{in}, \quad (30)$$

where Q_{abs} is the heat absorbed and Q_{in} is the heat input. The heat of combustion considered by the manufacturers varies from HHV, which typical in the US, to LHV, typical in EU, or even beyond, LHV, considering the acid dew point. Consequently, manufacturers may provide notably different efficiencies for products with similar capabilities. For instance, Eq. (30), is usually calculated by the LHV in Europe, hence, it is common to see condensing domestic boilers with 108–109% efficiency in the market.

Upon knowing the fuel composition and its measured or estimated HHV, its oxidation is calculated next as a net reaction to determine its oxidizer—which is air in the majority of the practical cases—requirement. The procedure is the same for complex fuels as well, however, it is shown only for methane here while wood combustion is presented in Sect. 4.5. The calculated molar masses are $M_{CH_4} = 1 \cdot 12 + 4 \cdot 1$ kg/kmol, $M_{CO_2} = 1 \cdot 12 + 2 \cdot 16$ kg/kmol, $M_{H_2O} = 2 \cdot 1 + 1 \cdot 16$ kg/kmol. Air can be assumed as 21% O_2 and 79% N_2, hence 1 kmol O_2 comes with 3.762 kmol N_2; $M_{O_2} = 2 \cdot 16$ kg/kmol, $M_{N_2} = 2 \cdot 14$ kg/kmol. By using the mole numbers in Eq. (22), the mass balance can be calculated

$$16 \, kg \, CH_4 + 64 \, kg \, O_2 + 210.7 \, kg \, N_2 \longrightarrow 44 \, kg \, CO_2 + 36 \, kg \, H_2O + 210.7 \, kg \, N_2$$
$$(31)$$

By dividing Eq. (31) by the mass of the fuel, i.e., by 16 kg, Eq. (32) is formed:

$$1 \, kg \, CH_4 + 17.17 \, kg \, air \longrightarrow 2.75 \, kg \, CO_2 + 2.25 \, kg \, H_2O + 13.17 \, kg \, N_2 \quad (32)$$

There are two important quantities calculated in Eq. (32): the theoretical air require-ment for the combustion of 1 kg CH_4, which is $\mu_{L_0} = 17.17$ kg/kg, and the resulting flue gas mass that is always 1 kg more when there is no ash, i.e., $\mu_{V_0} = 18.17$ kg/kg here. It is highly advised to check the sum of masses on both sides to avoid calculation mistakes. Since combustion rarely occurs at stoichiometric conditions, the available air mass to the required air mass for complete combustion is a crucial parameter. It is called equivalence ratio and is defined as

$$\lambda = \frac{\mu_L}{\mu_{L_0}}, \tag{33}$$

where μ_L is the actual air mass introduced with 1 kg fuel. This is the air-to-fuel equiv-alence ratio, used for boilers and internal combustion engines (this is why the sensor in the exhaust manifold is called lambda probe) while the researcher community and the gas turbine industry favor its reciprocal, the fuel-to-air equivalence ratio, ϕ.

The next quantity to be calculated is the adiabatic flame temperature, T_{ad}, which is a theoretical limit that cannot be exceeded in the case of homogeneous combustion. Firstly, a heat balance is written as

$$\Delta H_{comb} + H_{fuel} + H_{air} = H_{fg}, \tag{34}$$

where ΔH_{comb} is the enthalpy released by combustion while the other terms are the enthalpy of fuel, air, and flue gas in order. Equation (34) can be used in general since it is an energy balance following the first law of thermodynamics. Similarly, the entropy production of combustion can also be calculated [218]. Nevertheless, this latter quantity is rarely required in simplified calculations of real combustion chambers, hence, it is omitted.[2] Assuming a steady process with a continuous air and fuel inlets, enough oxidizer to burn all the fuel, i.e., $\lambda > 1$, the time derivative of enthalpies in Eq. (34) can be calculated as

$$
\begin{aligned}
\partial_t \left(\Delta H_{comb} \right) &= m_{fuel} \cdot HHV, \\
\partial_t H_{fuel} &= m_{fuel} \cdot \bar{c}_{p,fuel} \cdot \left(T_{fuel} - T_0 \right), \\
\partial_t H_{air} &= m_{fuel} \cdot \lambda \cdot \mu_{L_0} \cdot \bar{c}_{p,air} \cdot \left(T_{air} - T_0 \right), \\
\partial_t H_{fg} &= m_{fuel} \cdot \left[\mu_{V_0} + (\lambda - 1) \cdot \mu_{L_0} \right] \cdot \bar{c}_{p,fg} \cdot \left(T_{ad} - T_0 \right),
\end{aligned}
\tag{35}
$$

---

[2]Beyond entropy, another environment-centric evaluation method, the exergy analysis reveals the effect of combustion in greater detail. Its single message is to use the appropriate heat source for a given process [219, 220], i.e., it is highly destructive to produce warm water ($\sim$40–60 °C) at home by using a natural gas-fired boiler that produces $\sim$1000 °C flue gas.

where m denotes the mass flow rate, $\bar{c}_p$ is the integral average of the specific heat at constant pressure, and T_0 is the reference temperature. By putting Eq. (35) to Eq. (34), T_{ad} is

$$T_{ad} = \frac{HHV + \bar{c}_{p,fuel} \cdot (T_{fuel} - T_0) + \lambda \cdot \mu_{L_0} \cdot \bar{c}_{p,air} \cdot (T_{air} - T_0)}{[\mu_{V_0} + (\lambda - 1) \cdot \mu_{L_0}] \cdot \bar{c}_{p,fg}} + T_0. \quad (36)$$

Equation (36) is independent of the fuel mass flow rate. If both the fuel and air temperatures are equal to the reference temperature, the nominator is simplified to HHV only. Calculation of $\bar{c}_{p,fg}$ usually requires a few iteration steps since this variable always show a notable change between the reference and the adiabatic flame temperatures. Conditions not considered:

- heat transfer between the flame and the surroundings—this is the cause of the name of T_{ad},
- detailed reactions,
- dissociation of species,
- spatial distribution of the mixture inside the combustion chamber.

Hence, the local flame temperature can exceed the adiabatic one in numerical simulations. Only the mass-weighted average temperature should be equal to or lower than T_{ad}. Excessive flame temperatures lead to dissociation and incomplete combustion as the chemical equilibrium concentrations start to shift towards the reactants from the products side. The most notable pollutant which concentration correlates exponentially with T_{ad} is NO_X [209]. Hence, thermal power plants are designed to 1500 K maximum flame temperature to avoid the excessive formation of this pollutant. Due to the associated longer time scales, only quick quench can stop nitrogen chemistry before it exploits the high flame temperature [173]. Practical outcomes of Eq. (36) which affect T_{ad} are the following.

- The heating value of the fuel. Lower HHV decreases its value, hence, to keep it the same, other parameters have to be adjusted.
- Both fuel and oxidizer preheating increases T_{ad}. Boilers designed for high efficiency contain air preheater in the lower price segment and fuel preheater—if applicable—beyond that. When the boiler is used in a thermal power plant for steam generation, the feedwater preheater, called economizer, is applied before the other two. Appropriate heat exchangers hence can utilize the remaining heat of the flue gas upon leaving the combustion chamber, and only the above-mentioned acid dew point sets a real limitation to the maximum heat extraction. Since the natural gas is usually free of sulfur, domestic condensing boilers condense a significant portion of the H_2O content of the flue gas, utilizing almost all the HHV of the fuel.
- Equivalence ratio. To reduce the average combustion temperature, air dilution is commonly used in practice. However, even in steady combustion, the theoretical limitation, which is ca. $\lambda = 2$ [203] for hydrocarbon flames, unsteady phenomena start to rise when this value is approached [221–223].

• Flue gas recirculation and oxy-fuel combustion. The former one increases μ_{L_0} while the latter one decreases it and T_{ad} as well.

There is one field where adiabatic conditions are highly desired which is laminar flame speed research [224–226]. Even though it has more than 150 years of history, the development of advanced numeric codes demands precise laminar flame speed values since all the turbulent flame speed calculations are based on that. Even in the past four decades, papers claimed to be accurate, measured it between 33 and 43 m/s for methane and air at 300 K and 1 atm conditions. By the years, there is no clear convergence towards a well-defined value. Considering only the experiments of the last decade, the limits decreased to 34 and 40 m/s which still means a considerable measurement uncertainty that affects the numerical models and the results calculated by them.

The above quantities and phenomena enable now the classical thermal analysis to be performed. Namely, the calculation of the thermal interactions between the flame and its ambient: convection and thermal radiation. Expanding the left hand side of Eq. (34) gives:

$$H_{fg} = H_h + H_r + H_{loss}, \tag{37}$$

where subscripts h, r, and loss refer to convection, thermal radiation, and flue gas loss, respectively. The cause of the last term is already discussed in the second item of the above list. There are physical limits which do not allow the full extraction of the available enthalpy of the flue gas. Hence, the exhaust gas temperature will be at least several degrees higher than the ambient temperature; it cannot be avoided. Usually, the measurement of losses is more reasonable as they are smaller quantities than the used enthalpies. By using a sensor with an uncertainty of a few percents, it is an obvious conclusion that the smaller quantities lead to lower measurement uncertainty in Eq. (37).

Convection in combustion chambers can be calculated by using Newton's law of cooling; no additional theoretical correction is required:

$$q_h = h \cdot (T_{fg} - T_w), \tag{38}$$

where q_h is the heat flux by convection, h is the heat transfer coefficient, and T_w is the wall temperature. More details on designing and calculating heat exchangers are available in Chap. 7 of Ref. [172]. However, T_{fg} is too high in gas turbines to let the flue gas touching the wall. Therefore, various cooling techniques are available to use the compressor bleed air to form a thin protective boundary layer above the surface of the combustion chamber, see Chap. 8 in Ref. [173]. Since the mixing of fluids with various viscosities is slow, this solution prevents the flue gas from touching any solid part of the chamber. In addition, the next 1–2 turbine stages are also cooled, therefore, the flue gas touches metal only when its temperature is notably reduced through losing enthalpy by driving the turbines.

The thermal radiation can be calculated by Eq. (39):

Fig. 23 Effect of droplet size on the flame color

$$q_r = \varepsilon_{flame,abs} \cdot \varphi \cdot \sigma_0 \cdot \left(T_{flame}^4 - T_{abs}^4\right), \tag{39}$$

where $\varepsilon_{flame,abs}$ is the common emissivity of the flame and the absorbing material, φ is the view factor, and σ_0 is the Stefan-Boltzmann constant. Subscripts *flame* and *abs* refer to the flame and the absorbing material, respectively.

The heat transfer by radiation, excluding σ_0, contains only non-trivial variables. The flame temperature varies spatially, and the determination of the full temperature field requires Computational Fluid Dynamics (CFD) calculations. However, there is a rough estimation, used in boiler design, which may serve as a first guess:

$$T_{flame} = \sqrt{T_{ad} \cdot T_{fg}}, \tag{40}$$

where T_{fg} is the flue gas temperature that leaves the combustion chamber. Assuming that all the other parameters are known, and there is heat transfer inside the chamber, this step alone requires iterations. Since it is a highly nonlinear equation to solve due to T^4, a suitable numerical solver algorithm is necessary.

T_{abs} notes all materials which interact in the radiative heat exchange. Besides the solid surfaces, ash and triatomic gases (CO_2, H_2O, and SO_2) also participate in the heat transfer via radiation. Therefore, the combustion of all practical fuels is affected. The simplest radiation model for this purpose in numerical codes is the P-1 [214]. All of the radiation models already contain assumptions which can be directly used for calculating the view factors. In non-premixed flames, the flame surface can be used for radiative heat transfer calculations while premixed flames, governed by volumetric reactions, are modeled as hot gases.

There are typical emissivity values present in the literature for various fuels, equivalence ratios, combustion chamber designs. Section 6.2 of Ref. [172] provides a good overview of the emissivity of various flames. To demonstrate the complexity of the determination of ε, Fig. 23 shows two images of diesel oil combustion at two atomizer pressures while the other parameters are closely the same.

The flame on the left is the result of larger fuel droplet sizes as the atomizing pressure was slightly lowered. Hence, even though the global equivalence ratio is identical, the incomplete evaporation of the droplets lead to non-premixed combustion locally. In this case, soot particles are formed which behave as black body emitters. The soot concentration drastically modifies the overall radiative heat transfer. This is the reason why industrial boilers with radiative heat exchangers feature bright flame while it is a drawback in gas turbine combustion. Note that the emissivity is affected by the oxidizer and operating pressure besides combustion chamber design.

The flame color is determined by the concentration of the excited intermediate species. The blue light is emitted by the CH* at principally 431 nm, while the greenish color is slightly visible above the stoichiometric concentration. This is emitted by the C_2 radicals at 516 nm. OH* has a strong emission in the range between 300 and 310 nm which is in the UV regime, hence, it is invisible. This is the reason why flame detectors focus on this radical. Typical practical flame temperatures lie in the range of 1000–2000 °C where the majority of the emitted radiation is located in the infrared regime. More information on the emission of various species can be found in Ref. [210].

4.5 Example: Wood-Fired Combustion Chamber

The following example will present the calculation of a wood-fired 200 kW combustion chamber which can be part of a boiler or other technological device. The goal is the determination of the adiabatic flame temperature of combustion with air, then with flue gas recirculation, and finally when oxy-fuel combustion is applied. Table 5 contains the composition of the wood, being the basis of this example. Substituting the composition into Eq. (28):

$$HHV = 0.3491 \cdot 50 + 1.1783 \cdot 6 - 0.1034 \cdot 41 - 0.0151 \cdot 1 - 0.0211 \cdot 2 = 20.23 \, \text{MJ/kg}.$$
(41)

The consumption rate of the wood at nominal thermal power:

$$m_{wood} = \frac{200 \, \text{kW}}{20.23 \, \text{MJ/kg}} = 9.89 \, \text{g/s} = 35.6 \, \text{kg/h}.$$
(42)

Table 5 Mass composition of the wood used for the calculation [227]

Carbon (C)	50
Oxygen (O)	41
Hydrogen (H)	6
Nitrogen (N)	1
Ash	2

The next step is the calculation of the oxygen mass required to combust 1 kg wood, $\mu_{O_2,req}$. Knowing that 1 kg C requires 32/12 kg O_2 to form CO_2 and 1 kg H_2 requires 16/2 kg O_2 to form H_2O:

$$\mu_{O_2,req} = 0.5 \cdot \frac{32}{12} + 0.06 \cdot \frac{16}{2} - 0.41 = 1.4 \; \frac{kg}{kg\ wood}. \tag{43}$$

This value will be used later for oxy-fuel combustion. The required air mass, μ_{L_0} for the combustion of 1 kg wood is

$$\mu_{L_0} = \mu_{O_2,req} \cdot \left(\frac{79}{21} \cdot \frac{28}{32} + 1\right) = 6.02 \; \frac{kg\ air}{kg\ wood}. \tag{44}$$

The specific flue gas mass to 1 kg fuel, μ_{V_0}, will not be simply $\mu_{L_0}+1$ kg/kg as the wood here contains 2% ash. In addition, it is necessary to know the mass percentage of bottom ash in the total ash content, r_b, to continue with the calculations. The fly ash will leave the chamber with the flue gas. Considering a clean technology, $r_b = 1$ is used here. Hence, μ_{V_0} is the following:

$$\mu_{V_0} = \mu_{L_0} + 1 - a \cdot r_b = 7 \; \frac{kg\ fg}{kg\ wood}. \tag{45}$$

The bottom ash has been disappeared from the stoichiometric flue gas mass. The stoichiometric net reaction with masses specific to 1 kg wood:

$$0.5\ kg\ C + 0.41\ kg\ O + 0.06\ kg\ H + 0.01\ kg\ N + 0.02\ kg\ a + 1.4\ kg\ O_2 + 4.62\ kg\ N_2$$
$$\longrightarrow 1.83\ kg\ CO_2 + 0.54\ kg\ H_2O + 4.63\ kg\ N_2 + 0.02\ kg\ a. \tag{46}$$

After finishing with the necessary stoichiometric calculations, the thermal properties of the species are required. To facilitate complete combustion of wood, air-to-fuel equivalence ratio, λ, was selected as 1.6 [228]. The inlet thermodynamic data is listed in Table 6.

As the reference temperature is equal to that of the air inlet and wood inlet, Eq. (36) simplifies to the following form, considering the ash content,

$$T_{ad} = \frac{HHV}{\left[\mu_{V_0} + (\lambda - 1) \cdot \mu_{L_0} + (1 - r_b) \cdot a\right] \cdot \bar{c}_{p,fg}} + T_0 = 1456\,^{\circ}C. \tag{47}$$

Table 6 Thermodynamic data of the inlet

T_0	15	°C
T_{wood}	15	°C
T_{air}	15	°C
$T_{p,wood}$	1.5	kJ/(kg·K)
$c_{p,air}$	1	kJ/(kg·K)

Fig. 24 Mass flows of the combustion chamber with flue gas recirculation

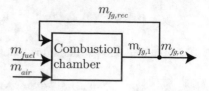

$\bar{c}_{p,fg}$ was calculated through three iteration steps with a final value of 1.322 kJ/(kg·K). As $r_b = 1$, the term containing the ash is zero since the ash remaining in the combustion chamber. In reality, it is not feasible, therefore, the technique which is responsible for ash removal should be considered in the energy balance. Currently, this is omitted. The flue gas mass flow rate can be calculated as:

$$m_{fg} = m_{wood} \cdot \left[\mu_{V_0} + (\lambda - 1) \cdot \mu_{L_0} + (1 - r_b) \cdot a \right] = 0.105 \text{ kg/s} = 378 \text{ kg/h}. \tag{48}$$

This flow rate is the basis of designing a chimney which is discussed elsewhere [171].

To reduce the flue gas temperature, flue gas recirculation can be applied, shown in Fig. 24. In the current example, recirculation rate, $r_{rec} = m_{fg,rec} / m_{fg,o} = 0.5$ is assumed.

In this case, Eq. (34) is supplemented by an additional term, $H_{fg,rec}$. Considering the recirculating flue gas temperature as 150 °C—assuming that the combustion chamber belongs to a boiler that is designed to operate also on coal, hence, the acid dew point is not reached. Presently, Eq. (48) will be equal to $m_{fg,o}$, while $m_{fg,1} = m_{fg,o} \cdot (1 + r_{rec})$. Hence, the new adiabatic flame temperature, $T_{ad,rec}$, will be the following:

$$T_{ad,rec} = \frac{m_{wood} \cdot HHV + r_{rec} \cdot m_{fg,o} \cdot \bar{c}_{p,fg,rec} \cdot (T_{rec} - T_0)}{m_{wood} \cdot (1 + r_{rec}) \left[\mu_{V_0} + (\lambda - 1) \cdot \mu_{L_0} \right] \cdot \bar{c}_{p,fg,rec}} + T_0 = 1056 \text{ °C}, \quad (49)$$

where $\bar{c}_{p,fg,rec} = 1.275$ kJ/(kg·K) was resulted at the end of the iterations.

The adiabatic flame temperature will be hence lower while the combustion power and the fuel remain the same. However, the equivalence ratio in the combustion chamber will increase to $\lambda_{rec} = \lambda + (\lambda - 1) \cdot r_{rec} = 1.9$. If one principal aim is to keep the equivalence ratio at the same, then the air mass flow rate should be adjusted to the new conditions, i.e., $\lambda_{0,rec} = 1.4$. By repeating the above calculation, including the used results of the case without flue gas recirculation, the updated adiabatic flame temperature will be 1162 °C at $\bar{c}_{p,fg,rec} = 1.3$ kJ/(kg·K). The new adiabatic flame temperature is slightly higher, following the expectations, however, its value is well below T_{ad}.

In the followings, oxy-fuel combustion is calculated. Recalling Eq. (43), $\mu_{O_2,req} = 1.4$ (kg O_2)/(kg wood), the specific flue gas mass for 1 kg wood, considering the oxygen-to-fuel equivalence ratio, will be the following:

$$\mu_{V,oxy} = \mu_{O_2,req} \cdot \lambda + 1 - a \cdot r_b + a \cdot (1 - r_b) = 3.23 \, \frac{\text{kg fg}}{\text{kg wood}}. \tag{50}$$

The resulting flue gas mass flow rate

$$m_{fg,oxy} = m_{wood} \cdot \mu_{V,oxy} = 31.9 \, \text{g/s} = 115 \, \text{kg/h}. \tag{51}$$

Hence, the adiabatic flame temperature is

$$T_{ad,oxy} = \frac{HHV}{\mu_{V,oxy} \cdot \bar{c}_{p\,fg,oxy}} + T_0 = 3719 \, ^{\circ}\text{C}. \tag{52}$$

It can be seen that this high adiabatic flame temperature is highly impractical since the combustion chamber will be too expensive if it is even possible to manufacture one from the available materials for such conditions. Therefore, flue gas recirculation is necessary again to end up with reasonable combustion chamber temperature, aiming the range of 1000–1200 °C.

Continuing with a recirculation rate, $r_{oxy,rec} = 3$ which will lead to a similar flue gas flow rate from the combustion chamber than in the previous case with recirculation. Again, let's use $T_{oxy,rec} = 150$ °C. The mass flow rate leaving the combustion chamber is

$$m_{fg,oxy,rec} = \left(1 + r_{oxy,rec}\right) m_{wood} \cdot \mu_{V,oxy} = 0.128 \, \text{kg/s} = 459 \, \text{kg/h}. \tag{53}$$

The resulting adiabatic flame temperature with $\bar{c}_{p\,fg,oxy,rec} = 1.379$ kJ/(kg·K):

$$T_{ad,oxy,rec} = \frac{m_{wood} \cdot HHV + m_{wood} \cdot \mu_{V,oxy} \cdot \bar{c}_{p\,fg,oxy,rec} \cdot \left(T_{oxy,rec} - T_0\right)}{m_{fg,oxy,rec} \cdot \bar{c}_{p\,fg,oxy,rec}} + T_0 = 1162 \, ^{\circ}\text{C}. \tag{54}$$

To adjust the results to $\lambda_{oxy} = 1.6$, the iterated $\bar{c}_{p\,fg,oxy,rec}$ value was 1.745 kJ/(kg·K), leading to an adiabatic flame temperature of 1270 °C. However, the high oxygen concentration suggests that combustion will be complete at $\lambda = 1.9$ as well.

As a short summary for the present example, it is clear that either flameless or oxy-fuel combustion is desired, the flue gas recirculation must be solved as both technologies demand it. Designing a combustion chamber for flameless combustion in gas turbines is presently a hot topic [229–231].

5   Building Blocks of Thermal Power Cycles

All the thermal power cycles used currently in large scales feature four principal processes:

- compressing the working fluid (the compressor makes work on it),

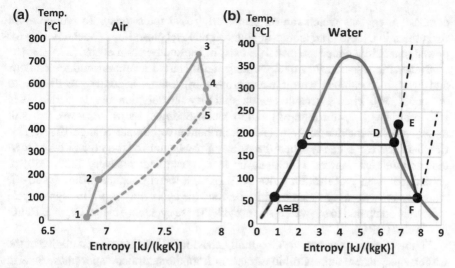

Fig. 25 Classical thermal cycles. **a** Joule cycle, **b** Rankine cycle [232]. With permission

- introducing heat,
- expansion of the working fluid (it makes work on the expansion device),
- removing the heat to return to the initial state.

Figure 25 shows two examples: the Joule cycle with air, and the Rankine cycle with water as working fluids. The reason of the use of four processes is the following. During compression (1 → 2 in Fig. 25a, A → B in Fig. 25b), the work to be made is smaller than the work available through expansion (3 → 5 in Fig. 25a, E → F in Fig. 25b), after heat is introduced.

It is possible that new concepts will emerge in thermal energy use in the future, however, the building blocks are likely to remain similar. Therefore, a few properties of these elementary processes are discussed in the present section. It starts with working fluids in Sect. 5.1, then compressors and expansion devices are discussed in Sect. 5.2. Thirdly, heat exchangers are detailed in Sect. 5.3, also focusing on small-volume devices, like microelectronic components, where the heat generation is in the range of $10 \, W/mm^3$. Lastly, thermoelectric applications are discussed in Sect. 5.4 as they are simple and small devices for either heating or cooling or generate electricity from a temperature difference. For a more detailed discussion on thermal cycles, see Ref. [218].

5.1 Working Fluids

Open cycles will remain using air as working fluid since even the cost of water treatment is too high to release it after expansion. These cycles practically mean combustion engines which run on principally fossil fuels today. To reduce fossil fuel utilization, there is an increasing interest in novel closed-cycle technologies for power generation, which are based on existing thermodynamic cycles [233–235].

Systematic research for novel working fluids started when chlorofluorocarbons and hydrofluorocarbons were banned from refrigerators, according to the Kyoto protocol [236], when it became evident that they severely damage the ozone layer. R134a became the substitute of them in most cooling systems, however, it is still under replacement due to its notable contribution to climate change [237]. This rich database came handy when the research has begun for the utilization of low-temperature heat sources which can be, e.g., waste heat recovery [238, 239] or geothermal energy utilization [240–242]. Here, low-temperature means <400 °C, where organic working fluids used in the Rankine cycle may result in higher cycle efficiency compared to the water–steam cycle. These cycles are called Organic Rankine Cycles (ORC).

Figure 26 shows various working fluid behaviors. Water is a representative of the first category where an isentropic expansion from the saturated vapor phase ends in the wet regime. Figure 26b shows a dry fluid where the isentropic expansion ends in the superheated regime. It is advantageous from one side since the appearing droplets in wet fluids erodes the expansion devices [243], and this fluid type is free from that phenomenon. Nevertheless, the cycle efficiency will be low if the end of the expansion is too far from the saturation line. The third example, the isentropic fluid does not exist, however, it would be characterized by the highest possible cycle efficiency for the application. More information on working fluids, their thermodynamic properties, and classification is published by the research group of Imre [244–247].

5.2 Compressors and Expansion Devices

The operation of all compressors and expansion devices can be estimated by using the first law of thermodynamics, regardless of the design. Therefore, all the initial thermal calculations can be performed, and a process/cycle can be analyzed without the need for geometric parameters or datasheets. This is emphasized in this subsection. Then few references are given to give an overview of waste heat utilization of existing thermal power cycles and industrial processes. Even though there are several attractive solutions present, financial and practical aspects may overwrite them if the gain would not worth the effort. These decisions are always made for the individual system and may vary by region even for a similar device due to, e.g., governmental support of certain technologies.

The power required to drive a compressor, regardless of the device design, can be estimated by Eq. (55)

$$P_C = m_C \cdot (h_2 - h_1), \tag{55}$$

where m is the mass flow rate of the medium and h is the specific enthalpy. For convention, compression processes are denoted with $1 \rightarrow 2$, and $3 \rightarrow 4$ is used for

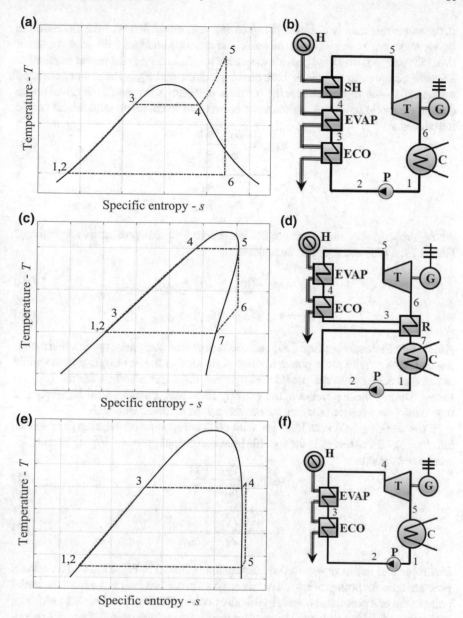

Fig. 26 Cycles for various working fluid types for ORC applications. **a** Wet, **b** dry, and **c** isentropic working fluids [244]. With permission

expansion in this subsection. Hence, the power delivered by the expansion device is

$$P_E = m_E \cdot (h_3 - h_4).$$ (56)

If the same medium is used for compression and expansion and the operation is steady-state, $m_C = m_E = m$ can be used. The enthalpy changes have been chosen in Eqs. (55) and (56) to always provide a positive result, i.e., P_C is the power required to drive the compressor and make work on the medium, and P_E is the power generated by the expansion device, hence, the medium performs work on the device. During design, adiabatic processes are the most convenient to calculate with which can be introduced as

$$\eta_{ad,C} = \frac{h_{2,is} - h_1}{h_{2,re} - h_1}, \tag{57}$$

$$\eta_{ad,E} = \frac{h_3 - h_{4,re}}{h_3 - h_{4,is}}, \tag{58}$$

where subscripts is and re denote the isentropic and real processes, respectively. Hence, Eqs. (55) and (56) can be updated as

$$P_C = m_C \cdot \left(h_{2,is} - h_1\right) / \eta_{ad,C}, \tag{59}$$

$$P_E = m_E \cdot \left(h_3 - h_{4,is}\right) \cdot \eta_{ad,E}. \tag{60}$$

Adiabatic efficiencies in Eqs. (59) and (60) show that compressors with thermodynamic losses require more power to drive compared to the isentropic process while expansion devices always provide less power if the process has thermodynamic losses. During these processes, the entropy increases, compared to the isentropic case, which is a consequence of the second law of thermodynamics.

If the working fluid is an ideal gas with constant specific heat during the process, i.e., $h = c_p \cdot T$, where c_p is the specific heat at constant pressure, Eqs. (57) and (58) can be rewritten as

$$\eta_{ad,C} = \frac{T_{2,is} - T_1}{T_{2,re} - T_1}, \tag{61}$$

$$\eta_{ad,E} = \frac{T_3 - T_{4,re}}{T_3 - T_{4,is}}. \tag{62}$$

Isentropic and real compression and expansion is shown in Fig. 27, using a constant pressure ratio for both processes. Even though all processes are drawn from the initial point to the end point, real ones are typically non-equilibrium processes. In addition, the working fluid is not stopped in a real application to achieve steady-state, therefore, the measurement location also affects the result of the indication procedure. Hence, all these thermodynamic diagrams bear notable uncertainties when one wishes to illustrate a real process.

Using the adiabatic relations, the pressure and volume ratios can be expressed with the temperature ratio

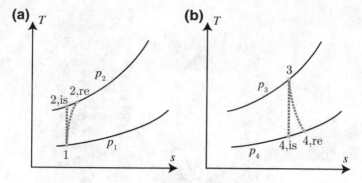

Fig. 27 Isentropic and real **a** compression, **b** expansion

$$\frac{T_2}{T_1} = \left(\frac{p_2}{p_1}\right)^{\frac{\gamma-1}{\gamma}} = \left(\frac{V_2}{V_1}\right)^{1-\gamma}, \qquad (63)$$

where γ is the ratio of specific heats or adiabatic exponent, p is the pressure, and V is the volume. The reason for discussion of both is that dynamic compressors discuss pressure ratio, r_p, while the datasheet of positive displacement compressors contains the volume ratio, r_V. Since the appropriate ratio and the initial temperature of the process is usually known, Eqs. (59) and (60) can be rearranged as

$$P_C = m_C \cdot c_{p,C} \cdot T_1 \left(r_p^{\frac{\gamma-1}{\gamma}} - 1\right)/\eta_{ad,C}, \qquad (64)$$

$$P_E = m_E \cdot c_{p,C} \cdot T_3 \left(1 - 1/r_p^{\frac{\gamma-1}{\gamma}}\right) \cdot \eta_{ad,E}. \qquad (65)$$

When $P_C = P_E$, the operation is idle, i.e., there is no shaft power produced. This is roughly the case of jet engines, where the turbine utilizes only a portion of the available enthalpy which covers the power requirement of the compressor. Then the remaining enthalpy is converted into kinetic energy that produces the thrust. Other gas turbine layouts feature two separate turbine sections: one drives the compressor, and they are called gas generator together. The second section may drive a generator or a propeller in the case of helicopters and turboprop engines. This design allows more flexible operation of the system since the two shafts can run at different RPMs.

The device selection for a given application is usually based on the mass flow rate and adiabatic efficiency. This latter parameter contains all the thermodynamical and fluid dynamical losses. The expansion device types, suitable for a wide size range of expansion device powers, are shown in Fig. 28. The complete design procedure for an ORC system is discussed by Pethurajan et al. [238].

Waste heat recovery solutions with various unconventional bottoming cycles are evaluated by Saghafifar et al. [249] and Omar et al. [250]. Since internal combustion

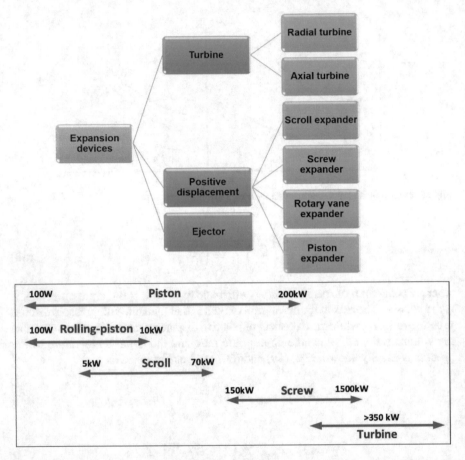

Fig. 28 Expansion device types (top) and their power range where they have high adiabatic efficiency (bottom) [248]. With permission

engines have around 40% efficiency, auxiliary thermal cycles can be used to increase this value [251–253]. Falter and Pitz-Paal [254] evaluated an efficient use of solar power to produce aviation-grade fuel. Concluding from the few cited, comprehensive papers, it is the challenge of the present engineer generation to drastically increase the thermal efficiency of our existing, high-performance devices. This is the other half besides direct renewable energy utilization to achieve a sustainable economy as it is likely that our current heat engines and power devices will remain in use for several additional decades.

5.3 *Heat Exchangers*

Closed thermodynamic cycles feature two heat exchanger systems; one for heat inlet and one for heat outlet. Nevertheless, they are common solutions when cooling or

Fig. 29 Heat transfer
between two separated fluids

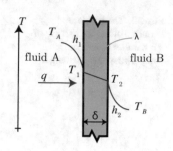

heating is required. Their dimension ranges from power plant size down to sub-millimeter, depending on the application. The discussion starts with fluid-fluid heat exchangers where either of them can be in a liquid or gaseous phase. This part follows the logic of Ref. [94] with few modification, and showing only the final formulas. Evaporation/condensation will be discussed later in this subsection.

Figure 29 shows the heat transfer between two, separated fluids under steady-state operation. This latter restriction will be used in the present subsection, however, it is fulfilled in the vast majority of the practical applications. In this case, the heat flow is calculated as

$$\dot{Q} = \frac{T_A - T_B}{1/(h_1 \cdot A) + \delta/(\lambda \cdot A) + 1/(h_2 \cdot A)}. \tag{66}$$

Note that the calculation of thermal resistances in Eq. (66) is true for planar surfaces. The formula for heat transfer coefficients of spherical and cylindrical bodies differs from those of planar problems and thermal conduction also differ. More details on them can be found in Refs. [94, 96]. In Eq. (66), the elementary heat transfer processes can be merged, and hence the overall heat transfer coefficient, U, can be used

$$\dot{Q} = U \cdot A \cdot (T_A - T_B). \tag{67}$$

Figure 30 shows the schematic temperature variation of parallel flow and counterflow heat exchangers as a function of the surface area. It highlights two problems with Eq. (67). It is usable only when T_A and T_B are given, and U is constant. In a real heat exchanger, all these parameters naturally vary along the fluid paths. To address these issues, Eq. (68) is used:

$$\dot{Q} = U \cdot f \cdot A \cdot \Delta T_m, \tag{68}$$

where f is a correction factor and is depending on the heat exchanger layout. This parameter is available in textbooks [94, 96] and vendors always provide this data, more often, a complete design manual. The other parameter, ΔT_m, is the logarithmic mean temperature difference which is

Fig. 30 Temperature
variation along the surface of
a **a** parallel flow **b**
counterflow heat exchanger

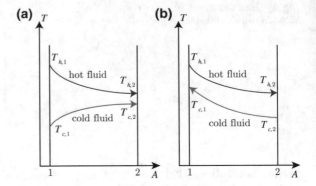

$$\Delta T_m = \frac{\left(T_{h,2} - T_{c,2}\right) - \left(T_{h,1} - T_{c,1}\right)}{\ln\left[\left(T_{h,2} - T_{c,2}\right) / \left(T_{h,1} - T_{c,1}\right)\right]}, \tag{69}$$

using the notations of Fig. 30. Equation (69) can be used for either parallel or counterflow heat exchangers. Note that solid parts from the fluids may form deposits on
the surfaces, decreasing U. Hence, either regular surface cleaning or fluid filtering
is necessary. More details on fouling and its mitigation can be found in the literature
[255–258].

The disadvantage of the presented method above is that all the temperatures have
to be known—or iterations required based on initial guess. In addition, if a heat
exchanger has to be selected for a given application, where all the parameters may
vary, an inverse method fits better. Therefore, the industry uses another approach,
called *NTU* or effectiveness method. *NTU* stand for Number of Transfer Units which
is calculated as

$$NTU = \frac{U \cdot A}{C_{\min}}, \tag{70}$$

where $C_{\min} = \min\left(m_h \cdot c_h, m_c \cdot c_c\right)$. m is the mass flow rate, and c is the heat capacity
of the fluid that is c_p for gases. The effectiveness is

$$\varepsilon_U = \frac{\dot{Q}}{\dot{Q}_{\max}} = \frac{\dot{Q}}{C_{\min} \cdot \left(T_{h,in} - T_{c,in}\right)}, \tag{71}$$

where $\dot{Q} = C_h \cdot \left(T_{h,in} - T_{h,out}\right) = C_c \cdot \left(T_{c,out} - T_{c,in}\right)$ is the actual heat flow, and
$\dot{Q}_{\max}$ is the maximum achievable heat flow. Subscripts in and out note inlet and
outlet, respectively. Generally, the effectiveness is $\varepsilon_U = \varepsilon_U \left(NTU, C_{\min}/C_{\max}\right)$. This
function can be derived by hand for the simplest heat exchanger cases, however, it is
an experimentally derived expression for advanced heat exchanger geometries which
are more common. Since the manufacturers provide all these functions, it is more
convenient to directly use them instead of making further assumptions and trying to
keep the generality.

The value of ε_U varies between 0 and 1, therefore, this quantity is often confused with efficiencies. Efficiency is the ratio of useful-to-inlet power, and the difference between them is the loss. However, the effectiveness of heat exchangers means how the heat exchange is related to the maximum available heat exchange—which would need an infinitely large surface area that is too expensive for any application. Only rejected heat means losses which decrease the efficiency; heat exchangers have inflows, outflows, and the heat content of the heating/cooling medium can be used further. Hence, ε_U has no direct relation to the efficiency of a device. Nevertheless, heat exchangers have a pressure drop which is a loss that cannot be avoided.

Heat exchangers, where phase change exists, usually separated to at least three units if all the subcooled liquid, superheated vapor, and evaporation/condensation occur during the process. For the former two parts, the heat exchange was already discussed. In the wet regime, where both liquid and vapor phases are present, the heat capacity is infinity. It is because any added heat will increase only the vapor content, the temperature of the mixture will remain the same while the fluid remains in the wet region. Theoretically, this consideration would enable the use of the above equations with $C_{max} = \infty$. Note that the phase change may lead to corrosion in practice which requires special attention [257, 260, 261]. The advantage of condensation is that the temperature remains constant, hence, such heat exchangers are extensively used in the food [262–264], pharmaceutical [265], and chemical industries [266] to provide a well-controlled thermal environment. If heat transfer is excessive during evaporation, dry-out may occur, i.e., all the liquid evaporates and the high incoming heat flux may damage the heat exchanger as the local heat transfer coefficient drops by several magnitudes. This is a common issue in boilers which generate superheated steam [39, 40].

A few compact heat exchanger types are shown in Fig. 31. These solutions provide a relatively large contact surface area in a small volume for efficient heat transfer between the fluids. The small size or low temperature difference in these units make the experimentally derived equations for larger sizes invalid. Consequently, designs emerge for compact heat exchangers which were considered ineffective in large sizes [267–269].

If there is a need for heat transport with high heat flux, heat pipes and vapor chambers can be used. These are common solutions in the cooling of microprocessors [271–273] and other electronic components with high power dissipation. In addition, they can be used for waste heat recovery and cooling purposes as well [274] besides space applications [275–277]. Its use in solar collectors is discussed in Sect. 2.1. Figure 32 shows the operation of a heat pipe. There is enhanced heat transfer in the evaporator and condenser stages, i.e., they contain fins or exposed to high/low temperature compared to that of the other side. The adiabatic section is either insulated or has a low surface area which negligibly affects the overall heat transfer. For more details on heat pipes and their design, see Ref. [278].

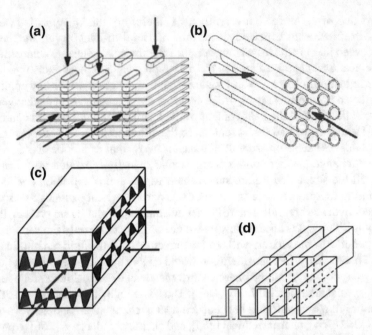

Fig. 31 Various compact heat exchangers. **a** Tube-fin, **b** tubular flow, **c** triangular fins, and **d** offset fin strips [259]. With permission

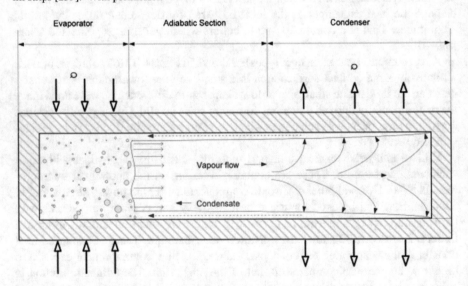

Fig. 32 Schematic diagram of the operation of heat pipes [270]. With permission

5.4 Thermoelectric Devices

Thermoelectric devices can be used for either heating or cooling. The Seebeck effect describes the direct electricity generation when two different conductors are exposed to a temperature difference between their junctions. The advantage of such devices is their small size, and they feature neither moving parts nor working fluid. In addition, they are able turn otherwise wasted heat into electricity [279] which makes such devices attractive to achieve an increased electric efficiency in various applications. Note that their size is limited by its thermal resistance, not by the manufacturing of the components. Nevertheless, classical thermodynamic power cycles can utilize heat with considerably higher efficiency, and the thermoelectric generators may degrade over time [280, 281]. Their design for optimum power output is discussed by Stevens [282] and Wang et al. [283]. Figure 33 shows how a thermoelectric generator works. The most significant obstacle of their use in large volume is that the thermoelectric elements have high electric an thermal conductivities. These material properties are closely related to each other. However, if there would be a material pair which would feature very low thermal and high electric conductivity, it would be an immediate success in waste heat recovery [284].

The inverse of the Seebeck effect was historically described by Peltier in 1834, i.e., when current is applied to the junction of two different conductors, a temperature difference will form which can be used for either cooling or heating. Again, applications favor them when its small size [285, 286], and no moving parts outweigh its low efficiency. Topology optimization was discussed by Lundgaard and Sigmund [287], who could increase the cooling performance of a unit by almost 50%. A potential application of this technology can be the cooling of the batteries of electric vehicles during charging [288].

In the case of thermoelectricity, the thermal balance needs to be supplemented by a charge balance equation. The thermoelectric effect can be calculated under steady-state conditions as [289]

$$q = -\lambda \frac{dT}{dx} + \mathcal{P} j_e, \tag{72}$$

where q is the heat flux, λ is the thermal conductivity, $\mathcal{P}$ is the Peltier coefficient, and j_e is the electric current density. When the last term is zero, it becomes the Fourier's law of heat conduction. The electric current density is

$$j_e = -\mathcal{S}\sigma \frac{dT}{dx} - \sigma \frac{dU_e}{dx}, \tag{73}$$

where $\mathcal{S}$ is the Seebeck coefficient, σ is the electrical conductivity, and U_e is the voltage. If there is no temperature gradient, Eq. (73) simplifies into Ohm's law. If either $\mathcal{S}$ or $\mathcal{P}$ becomes zero, the other term will be also zero. This can be seen from Eqs. (72) and (73). For detailed description on thermoelectricity from thermodynamical point of view, see Ref. [290].

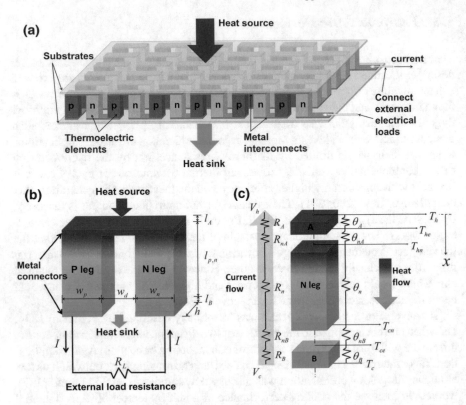

Fig. 33 **a** schematic model of a thermoelectric generator module. **b** fundamentals of a thermoelectric couple, **c** thermal resistance (θ) and electric resistance (R) network of one leg. Subscripts nA and nB refer to contact resistances between a leg and the metal connector [283]. With permission

References

1. U. Lohmann, J. Feichter, Global indirect aerosol effects: a review. Atmos. Chem. Phys. **5**, 715–737 (2005)
2. T. Wheeler, J. Von Braun, Climate change impacts on global food security. Science **341**(6145), 508–513 (2013)
3. M.P. McCarthy, M.J. Best, R.A. Betts, Climate change in cities due to global warming and urban effects. Geophys. Res. Lett. **37**(9), 1–5 (2010)
4. A. Dai, Drought under global warming: a review. Wiley Interdiscip. Rev.: Clim. Change **2**(1), 45–65 (2011)
5. J.A. Church, N.J. White, *A 20th Century Acceleration in Global Sea-Level Rise* (2006)
6. M.D. Flannigan, M.A. Krawchuk, W.J. De Groot, B.M. Wotton, L.M. Gowman, Implications of changing climate for global wildland fire. Int. J Wildland Fire, **18**(5), 483–507 (2009)
7. R.A. Pielke, C. Landsea, M. Mayfield, J. Laver, R. Pasch, Hurricanes and global warming. Bull. Am. Meteorol. Soc. **86**(11), 1571–1575 (2005)
8. W.R. Cline, *The Economics of Global Warming* (Peterson Institute for International Economics, Washington, D.C., 1992)
9. M.M. Mekonnen, P.W. Gerbens-Leenes, A. Y. Hoekstra, Future electricity: the challenge of reducing both carbon and water footprint. Sci. Total Environ. **569–570**, 1282–1288 (2016)

10. N.E. Vaughan, T.M. Lenton, A review of climate geoengineering proposals. Clim. Change **109**(3–4), 745–790 (2011)

11. A. Sandá, S.L. Moya, L. Valenzuela, Modelling and simulation tools for direct steam generation in parabolic-trough solar collectors: a review. Renew. Sustain. Energy Rev. **113**(June), 109226 (2019)

12. F. Meneguzzo, R. Ciriminna, L. Albanese, M. Pagliaro, The great solar boom: a global perspective into the far reaching impact of an unexpected energy revolution. Energy Sci. Eng. **3**(6), 499–509 (2015)

13. R. Ciriminna, F. Meneguzzo, M. Pecoraino, M. Pagliaro, Rethinking solar energy education on the dawn of the solar economy. Renew. Sustain. Energy Rev. **63**, 13–18 (2016)

14. Z. Dobrotkova, K. Surana, P. Audinet, The price of solar energy: comparing competitive auctions for utility-scale solar PV in developing countries. Energy Policy **118**(Jan), 133–148 (2018)

15. R.Y. Shum. Heliopolitics : the international political economy of solar supply chains. Energy Strateg. Rev. **26**(June 2019), 100390 (2020)

16. G.F. Nemet, E. O'Shaughnessy, R. Wiser, N. Darghouth, G. Barbose, K. Gillingham, V. Rai, Characteristics of low-priced solar PV systems in the U.S. Appl. Energy, **187**, 501–513 (2017)

17. J.-E. Zafrilla, G. Arce, M.-Á. Cadarso, C. Córcoles, N. Gómez, L.-A. López, F. Monsalve, M.-Á. Tobarra, Triple bottom line analysis of the Spanish solar photovoltaic sector: a footprint assessment. Renew. Sustain. Energy Rev. **114**(Feb), 109311 (2019)

18. M.J. De Wild-Scholten, Energy payback time and carbon footprint of commercial photovoltaic systems. Sol. Energy Mater. Sol. Cells **119**, 296–305 (2013)

19. S. Perry, J. Klemeš, I. Bulatov, Integrating waste and renewable energy to reduce the carbon footprint of locally integrated energy sectors. Energy **33**(10), 1489–1497 (2008)

20. V.M. Fthenakis, H.C. Kim, Greenhouse-gas emissions from solar electric- and nuclear power: a life-cycle study. Energy Policy **35**(4), 2549–2557 (2007)

21. V. Pranesh, R. Velraj, S. Christopher, V. Kumaresan, A 50 year review of basic and applied research in compound parabolic concentrating solar thermal collector for domestic and industrial applications. Sol. Energy **187**(Apr), 293–340 (2019)

22. L. Evangelisti, R. De Lieto Vollaro, F. Asdrubali, Latest advances on solar thermal collectors: a comprehensive review. Renew. Sustain. Energy Rev. **114**, 109318 (2019)

23. F. Bayrak, N. Abu-Hamdeh, K.A. Alnefaie, H.F. Öztop, A review on exergy analysis of solar electricity production. Renew. Sustain. Energy Rev. **74**(June 2016), 755–770 (2017)

24. H.A. Muhammed, S.A. Atrooshi, Modeling solar chimney for geometry optimization. Renew. Energy **138**, 212–223 (2019)

25. H.H. Al-Kayiem, O.C. Aja, Historic and recent progress in solar chimney power plant enhancing technologies. Renew. Sustain. Energy Rev. **58**, 1269–1292 (2016)

26. S. Akbarzadeh, M.S. Valipour, Heat transfer enhancement in parabolic trough collectors: a comprehensive review. Renew. Sustain. Energy Rev. **92**(Nov 2017), 198–218 (2018)

27. G.K. Manikandan, S. Iniyan, R. Goic, Enhancing the optical and thermal efficiency of a parabolic trough collector—a review. Appl. Energy **235**(Nov 2018), 1524–1540 (2019)

28. F.J. Collado, J. Guallar, Quick design of regular heliostat fields for commercial solar tower power plants. Energy **178**, 115–125 (2019)

29. A.A. Hachicha, B.A.A. Yousef, Z. Said, I. Rodríguez, A review study on the modeling of high-temperature solar thermal collector systems. Renew. Sustain. Energy Rev. **112**(June), 280–298 (2019)

30. G. Srilakshmi, N.S. Suresh, N.C. Thirumalai, M.A. Ramaswamy, Preliminary design of heliostat field and performance analysis of solar tower plants with thermal storage and hybridisation. Sustain. Energy Technol. Assess. **19**, 102–113 (2017)

31. M.J. Wagner, W.T. Hamilton, A. Newman, J. Dent, C. Diep, R. Braun, Optimizing dispatch for a concentrated solar power tower. Sol. Energy **174**(March), 1198–1211 (2018)

32. E. Bellos, C. Tzivanidis, A. Papadopoulos, Daily, monthly and yearly performance of a linear Fresnel reflector. Sol. Energy **173**(Nov 2017), 517–529 (2018)

33. E. Bellos, C. Tzivanidis, M.A. Moghimi, Reducing the optical end losses of a linear Fresnel reflector using novel techniques. Sol. Energy **186**(May), 247–256 (2019)
34. E. Bellos, Progress in the design and the applications of linear Fresnel reflectors—a critical review. Therm. Sci. Eng. Prog. **10**(Dec 2018), 112–137 (2019)
35. L. Sun, C. Zong, Y. Liang, W. Huang, Evaluation of solar brightness distribution models for performance simulation and optimization of solar dish. Energy **180**, 192–205 (2019)
36. R. Karimi, T.T. Gheinani, V.M. Avargani, A detailed mathematical model for thermal performance analysis of a cylindrical cavity receiver in a solar parabolic dish collector system. Renew. Energy **125**, 768–782 (2018)
37. S. Pavlovic, R. Loni, E. Bellos, D. Vasiljević, G. Najafi, A. Kasaeian, Comparative study of spiral and conical cavity receivers for a solar dish collector. Energy Convers. Manag. **178**(September), 111–122 (2018)
38. S. Ghazi, A. Sayigh, K. Ip, Dust effect on flat surfaces—a review paper. Renew. Sustain. Energy Rev. **33**, 742–751 (2014)
39. D.R.H. Jones, Creep failures of overheated boiler, superheater and reformer tubes. Eng. Fail. Anal. **11**(6), 873–893 (2004)
40. A. Arjunwadkar, P. Basu, B. Acharya, A review of some operation and maintenance issues of CFBC boilers. Appl. Therm. Eng. **102**, 672–694 (2016)
41. A.L. Avila-Marin, J. Fernandez-Reche, A. Martinez-Tarifa, Modelling strategies for porous structures as solar receivers in central receiver systems: a review. Renew. Sustain. Energy Rev. **111**(May), 15–33 (2019)
42. A. Schmitt, F. Dinter, C. Reichel, Computational fluid dynamics study to reduce heat losses at the receiver of a solar tower plant. Sol. Energy **190**(May), 286–300 (2019)
43. W.Q. Wang, Y. Qiu, Mi.J. Li, F. Cao, Z.B. Liu, Optical efficiency improvement of solar power tower by employing and optimizing novel fin-like receivers. Energy Convers. Manag. **184**(Dec 2018), 219–234 (2019)
44. S. Kiwan, A.L. Khammash, Investigations into the spiral distribution of the heliostat field in solar central tower system. Sol. Energy **164**(Feb), 25–37 (2018)
45. M. Atif, F.A. Al-Sulaiman, Optimization of heliostat field layout in solar central receiver systems on annual basis using differential evolution algorithm. Energy Convers. Manag. **95**, 1–9 (2015)
46. M.R. Rodríguez-Sánchez, A. Sánchez-González, D. Santana, Field-receiver model validation against Solar Two tests. Renew. Sustain. Energy Rev. **110**(May 2018), 43–52 (2019)
47. Lillyvilleky, *Aluminum Can Solar Heater* (2017)
48. Dr. Drashco, *DIY Solar Panels: The Ultimate Building Guide* (2019)
49. S. Karki, K.R. Haapala, B.M. Fronk, Technical and economic feasibility of solar flat-plate collector thermal energy systems for small and medium manufacturers. Appl. Energy **254**(July), 113649 (2019)
50. G. Faure, M. Vallée, C. Paulus, T.Q. Tran, Impact of faults on the efficiency curve of flat plate solar collectors: a numerical analysis. J. Clean. Prod. **231**, 794–804 (2019)
51. A. Fudholi, K. Sopian, A review of solar air flat plate collector for drying application. Renew. Sustain. Energy Rev. **102**(Dec 2018), 333–345 (2019)
52. V.V. Tyagi, N.L. Panwar, N.A. Rahim, R. Kothari, Review on solar air heating system with and without thermal energy storage system. Renew. Sustain. Energy Rev. **16**(4), 2289–2303 (2012)
53. S. Suman, M.K. Khan, M. Pathak, Performance enhancement of solar collectors—a review. Renew. Sustain. Energy Rev. **49**, 192–210 (2015)
54. R. Tang, W. Gao, Y. Yamei, H. Chen, Optimal tilt-angles of all-glass evacuated tube solar collectors. Energy **34**(9), 1387–1395 (2009)
55. E. Zambolin, D. Del Col, Experimental analysis of thermal performance of flat plate and evacuated tube solar collectors in stationary standard and daily conditions. Sol. Energy **84**(8), 1382–1396 (2010)
56. L.M. Ayompe, A. Duffy, M. Mc. Keever, M. Conlon, S.J. McCormack, Comparative field performance study of flat plate and heat pipe evacuated tube collectors (ETCs) for domestic water heating systems in a temperate climate. Energy **36**(5), 3370–3378 (2011)

57. A. Ibrahim, M.Y. Othman, M.H. Ruslan, S. Mat, K. Sopian, Recent advances in flat plate photovoltaic/thermal (PV/T) solar collectors. Renew. Sustain. Energy Rev. **15**(1), 352–365 (2011)

58. W. Pang, Y. Cui, Q. Zhang, Y. Hongwen, L. Zhang, H. Yan, Experimental effect of high mass flow rate and volume cooling on performance of a water-type PV/T collector. Sol. Energy **188**(June), 1360–1368 (2019)

59. M. Valizadeh, F. Sarhaddi, M. Adeli, Exergy performance assessment of a linear parabolic trough photovoltaic thermal collector. Renew. Energy **138**, 1028–1041 (2019)

60. M.A. Sharafeldin, G. Gróf, O. Mahian, Experimental study on the performance of a flat-plate collector using WO3/Water nanofluids. Energy **141**, 2436–2444 (2017)

61. M.A. Sharafeldin, G. Gróf, Evacuated tube solar collector performance using CeO2/water nanofluid. J. Clean. Prod. **185**, 347–356 (2018)

62. K. Farhana, K. Kadirgama, M.M. Rahman, D. Ramasamy, M.M. Noor, G. Najafi, M. Samykano, A.S.F. Mahamude, Improvement in the performance of solar collectors with nanofluids—a state-of-the-art review. Nano-Struct. Nano-Objects **18** (2019)

63. D. Wen, G. Lin, S. Vafaei, K. Zhang, Review of nanofluids for heat transfer applications. Particuology **7**(2), 141–150 (2009)

64. G. Colangelo, E. Favale, P. Miglietta, M. Milanese, A. de Risi, Thermal conductivity, viscosity and stability of Al2O3-diathermic oil nanofluids for solar energy systems. Energy **95**, 124–136 (2016)

65. M.E. Zayed, J. Zhao, Y. Du, A.E. Kabeel, S.M. Shalaby, Factors affecting the thermal performance of the flat plate solar collector using nanofluids: a review. Sol. Energy **182**(Nov 2018), 382–396 (2019)

66. J.H. Lee, K.S. Hwang, S.P. Jang, B.H. Lee, J.H. Kim, S.U.S. Choi, C.J. Choi, Effective viscosities and thermal conductivities of aqueous nanofluids containing low volume concentrations of Al2O3 nanoparticles. Int. J. Heat Mass Transf. **51**(11–12), 2651–2656 (2008)

67. R. Bubbico, G.P. Celata, F. D'Annibale, B. Mazzarotta, C. Menale, Experimental analysis of corrosion and erosion phenomena on metal surfaces by nanofluids. Chem. Eng. Res. Des. **104**, 605–614 (2015)

68. M. Grätzel, Photoelectrochemical cells. Nature **414**(6861), 338–344 (2001)

69. U. Pillai, Drivers of cost reduction in solar photovoltaics. Energy Econ. **50**, 286–293 (2015)

70. G. Kavlak, J. McNerney, J.E. Trancik, Evaluating the causes of cost reduction in photovoltaic modules. Energy Policy **123**(October), 700–710 (2018)

71. Y. Xu, J. Li, Q. Tan, A.L. Peters, C. Yang, Global status of recycling waste solar panels: a review. Waste Manag. **75**, 450–458 (2018)

72. B. Augustine, K. Remes, G.S. Lorite, J. Varghese, T. Fabritius, Recycling perovskite solar cells through inexpensive quality recovery and reuse of patterned indium tin oxide and substrates from expired devices by single solvent treatment. Sol. Energy Mater. Sol. Cells **194**(January), 74–82 (2019)

73. M. Fitra, I. Daut, M. Irwanto, N. Gomesh, Y.M. Irwan, TiO2 dye sensitized solar cells cathode using recycle battery. Energy Procedia **36**, 333–340 (2013)

74. A.M.K. Gustafsson, M.R.S.J. Foreman, C. Ekberg, Recycling of high purity selenium from CIGS solar cell waste materials. Waste Manag. **34**(10), 1775–1782 (2014)

75. W.H. Huang, W.J. Shin, L. Wang, W.C. Sun, M. Tao, Strategy and technology to recycle wafer-silicon solar modules. Sol. Energy **144**, 22–31 (2017)

76. P. Mandal, S. Sharma, Progress in plasmonic solar cell efficiency improvement: a status review. Renew Sustain. Energy Rev. **65**, 537–552 (2016)

77. M.K. Sahoo, P. Kale, Integration of silicon nanowires in solar cell structure for efficiency enhancement: a review. J. Materiomics **5**(1), 34–48 (2019)

78. D.G. Moon, S. Rehan, D.H. Yeon, S.M. Lee, S.J. Park, S.J. Ahn, Y.S. Cho, A review on binary metal sulfide heterojunction solar cells. Sol. Energy Mater. Sol Cells **200**(May), 109963 (2019)

79. M.S. Mozumder, A.H.I. Mourad, H. Pervez, R. Surkatti, Recent developments in multifunctional coatings for solar panel applications: a review. Sol. Energy Mater. Sol. Cells **189**(June 2018), 75–102 (2019)

80. Y. Galagan, E.W.C. Coenen, S. Sabik, H.H. Gorter, M. Barink, S.C. Veenstra, J.M. Kroon, R. Andriessen, P.W.M. Blom, Evaluation of ink-jet printed current collecting grids and busbars for ITO-free organic solar cells. Sol. Energy Mater. Sol. Cells **104**, 32–38 (2012)
81. B. Gerdes, M. Jehle, N. Lass, L. Riegger, A. Spribille, M. Linse, F. Clement, R. Zengerle, P. Koltay, Front side metallization of silicon solar cells by direct printing of molten metal. Sol. Energy Mater. Sol. Cells **180**(February), 83–90 (2018)
82. J.-M. Delgado-Sanchez, Solar Energy Materials and Solar Cells Luminescent solar concentrators: photo-stability analysis and long-term perspectives. Sol. Energy Mater. Sol. Cells **202**(July), 110134 (2019)
83. X. Cong, Z. Zhang, H. Yue, Y. Sheng, P. Jiang, H. Han, J. Zhang, Printed hole-conductor-free mesoscopic perovskite solar cells with excellent long-term stability using PEAI as an additive. J. Energy Chem. **27**(3), 764–768 (2018)
84. J. Maçaira, L. Andrade, A. Mendes, Laser sealed dye-sensitized solar cells: efficiency and long term stability. Sol. Energy Mater. Sol. Cells **157**, 134–138 (2016)
85. B. Wang, *First Commercial Perovskite Solar Late in 2019 and the Road to Moving the Energy Needle* (2019)
86. M.A. Mutalib, F. Aziz, A.F. Ismail, W.N.W. Salleh, N. Yusof, J. Jaafar, T. Soga, M.Z. Sahdan, N.A. Ludin, Towards high performance perovskite solar cells: a review of morphological control and HTM development. Appl. Mater. Today **13**, 69–82 (2018)
87. D.J. Friedman, Progress and challenges for next-generation high-efficiency multijunction solar cells. Curr. Opin. Solid State Mater. Sci. **14**(6), 131–138 (2010)
88. Q. Wali, N.K. Elumalai, Y. Iqbal, A. Uddin, R. Jose (2018) Tandem perovskite solar cells. Renew Sustain Energy Rev. **84**(Jan), 89–110 (2018)
89. J. Day, S. Senthilarasu, T.K. Mallick, Improving spectral modification for applications in solar cells: a review. Renew. Energy **132**, 186–205 (2019)
90. P. Singh, N.M. Ravindra, Temperature dependence of solar cell performance—an analysis. Sol. Energy Mater. Sol. Cells **101**, 36–45 (2012)
91. A. Lozano-Medina, L. Manzano, J.D. Marcos, A.M. Blanco-Marigorta, Design of a concentrating solar thermal collector installation for a hotel complex in Gran Canaria. Energy **183**, 803–811 (2019)
92. M. Ghorab, E. Entchev, L. Yang, Inclusive analysis and performance evaluation of solar domestic hot water system (a case study). Alex. Eng. J. **56**(2), 201–212 (2017)
93. S. Karki, K.R. Haapala, B.M. Fronk, Investigation of the combined efficiency of a solar/gas hybrid water heating system. Appl. Therm. Eng. **149**(Sept 2018), 1035–1043 (2019)
94. J.P. Holman (ed.), *Heat Transfer. McGraw-Hil:l Series in Mechanical Engineering*, 10th edn. (2009)
95. I. Subedi, T.J. Silverman, M.G. Deceglie, N.J. Podraza, Emissivity of solar cell cover glass calculated from infrared reflectance measurements. Sol. Energy Mater. Sol. Cells **190**(Sept 2018), 98–102 (2019)
96. M. Martin, K. Holge, *VDI Heat Atlas*, 2nd edn. (pringer, Berlin, 2010)
97. P. Hevia-Koch, J. Ladenburg, Where should wind energy be located? A review of preferences and visualisation approaches for wind turbine locations. Energy Res. Soc. Sci. **53**(February), 23–33 (2019)
98. M. Harper, B. Anderson, P.A.B. James, A.B.S. Bahaj, Onshore wind and the likelihood of planning acceptance: learning from a Great Britain context. Energy Policy **128**(Dec 2018), 954–966 (2019)
99. K. Kim, B. Song, V. Fernández-Hurtado, W. Lee, W. Jeong, L. Cui, D. Thompson, J. Feist, M.T.H. Reid, F.J. García-Vidal, J.C. Cuevas, E. Meyhofer, P. Reddy, Radiative heat transfer in the extreme near field. Nature **528**(7582), 387–391 (2015)
100. P. Scherhaufer, S. Höltinger, B. Salak, T. Schauppenlehner, J. Schmidt, A participatory integrated assessment of the social acceptance of wind energy. Energy Res. Soc. Sci. **45**(Nov 2017), 164–172 (2018)
101. E. Nordman, J. Mutinda, Biodiversity and wind energy in Kenya: revealing landscape and wind turbine perceptions in the world's wildlife capital. Energy Res. Soc. Sci. **19**, 108–118 (2016)

102. J.-S. Chou, O. Yu-Chen, K.-Y. Lin, Collapse mechanism and risk management of wind turbine tower in strong wind. J. Wind Eng. Ind. Aerodyn. **193**(July), 103962 (2019)
103. S.R. Brouwer, S.H.S. Al-Jibouri, I.C. Cárdenas, J.I.M. Halman, Towards analysing risks to public safety from wind turbines. Reliab. Eng. Syst. Saf. **180**(Nov 2016), 77–87 (2018)
104. F.O.M. Carneiro, H.H.B. Rocha, P.A.C. Rocha, Investigation of possible societal risk associated with wind power generation systems. Renew. Sustain. Energy Rev. **19**, 30–36 (2013)
105. S. Deshmukh, S. Bhattacharya, A. Jain, A.R. Paul, Wind turbine noise and its mitigation techniques: a review. Energy Procedia **160**(2018), 633–640 (2019)
106. S.S. Rodrigues, A.C. Marta, On addressing wind turbine noise with after-market shape blade add-ons. Renew Energy **140**, 602–614 (2019)
107. L. Fredianelli, S. Carpita, G. Licitra, A procedure for deriving wind turbine noise limits by taking into account annoyance. Sci. Total Environ. **648**, 728–736 (2019)
108. E.V. Bräuner, J.T. Jørgensen, A.K. Duun-Henriksen, C. Backalarz, J.E. Laursen, T.H. Pedersen, M.K. Simonsen, Z.J. Andersen, Long-term wind turbine noise exposure and the risk of incident atrial fibrillation in the Danish Nurse cohort. Environ Int. **130**(Mar), 104915 (2019)
109. J. Chen, F. Wang, K.A. Stelson, A mathematical approach to minimizing the cost of energy for large utility wind turbines. Appl. Energy **228**(June), 1413–1422 (2018)
110. D. Song, J. Liu, J. Yang, M. Su, S. Yang, X. Yang, Y.H. Joo, Multi-objective energy-cost design optimization for the variable-speed wind turbine at high-altitude sites. Energy Convers. Manag. **196**(Jan), 513–524 (2019)
111. D. Lande-Sudall, T. Stallard, P. Stansby, Co-located deployment of offshore wind turbines with tidal stream turbine arrays for improved cost of electricity generation. Renew. Sustain. Energy Rev. **104**(February), 492–503 (2019)
112. P. Enevoldsen, F.H. Permien, I. Bakhtaoui, A.K. von Krauland, M.Z. Jacobson, G. Xydis, B.K. Sovacool, S.V. Valentine, D. Luecht, G. Oxley, How much wind power potential does europe have? Examining european wind power potential with an enhanced socio-technical atlas. Energy Policy, **132**(Apr), 1092–1100 (2019)
113. D.S. Ryberg, D.G. Caglayan, S. Schmitt, J. Linßen, D. Stolten, M. Robinius, Detailed distribution and simulation of advanced turbine designs. The future of European onshore wind energy potential. Energy **182**, 1222–1238 (2019)
114. A. Bahrami, A. Teimourian, C.O. Okoye, H. Shiri, Technical and economic analysis of wind energy potential in Uzbekistan. J. Clean. Prod. **223**, 801–814 (2019)
115. A.M. Kaynia, Seismic considerations in design of offshore wind turbines. Soil Dyn. Earthq. Eng. **124**(Sept 2017), 399–407 (2019)
116. B. Yeter, Y. Garbatov, C.G. Soares, Risk-based life-cycle assessment of offshore wind turbine support structures accounting for economic constraints. Struct. Saf. **81**(June), 101867 (2019)
117. X. Wu, Y. Hu, Y. Li, J. Yang, L. Duan, T. Wang, T. Adcock, Z. Jiang, Z. Gao, Z. Lin, A. Borthwick, S. Liao, Foundations of offshore wind turbines: a review. Renew. Sustain. Energy Rev. **104**(Dec 2018), 379–393 (2019)
118. Y. Zhao, Z. Cheng, P.C. Sandvik, Z. Gao, T. Moan, An integrated dynamic analysis method for simulating installation of single blades for wind turbines. Ocean Eng. **152**(7491), 72–88 (2018)
119. Z. Ren, R. Skjetne, Z. Jiang, Z. Gao, A.S. Verma, Integrated GNSS/IMU hub motion estimator for offshore wind turbine blade installation. Mech. Syst. Signal Process. **123**, 222–243 (2019)
120. B.R. Sarker, T.I. Faiz, Minimizing transportation and installation costs for turbines in offshore wind farms. Renew. Energy **101**, 667–679 (2017)
121. A.Z. Dhunny, Z. Allam, D. Lobine, M.R. Lollchund, Sustainable renewable energy planning and wind farming optimization from a biodiversity perspective. Energy **185**, 1282–1297 (2019)
122. D. Ferreira, C. Freixo, J.A. Cabral, M. Santos, Is wind energy increasing the impact of socio-ecological change on Mediterranean mountain ecosystems? Insights from a modelling study relating wind power boost options with a declining species. J. Environ. Manag. **238**(Feb), 283–295 (2019)

123. K. Barré, I. Le Viol, Y. Bas, R. Julliard, C. Kerbiriou, Estimating habitat loss due to wind turbine avoidance by bats: implications for European siting guidance. Biolog. Conserv. **226**(August), 205–214 (2018)
124. K.F. Forbes, E.M. Zampelli, Wind energy, the price of carbon allowances, and CO_2 emissions: evidence from Ireland. Energy Policy **133**(July), 110871 (2019)
125. E. Rusu, F. Onea, An assessment of the wind and wave power potential in the island environment. Energy **175**, 830–846 (2019)
126. M. Veigas, G. Iglesias, Wave and offshore wind potential for the island of Tenerife. Energy Convers. Manag. **76**, 738–745 (2013)
127. E.G. Sakka, D.V. Bilionis, D. Vamvatsikos, C.J. Gantes, Onshore wind farm siting prioritization based on investment profitability for Greece. Renew. Energy (2019)
128. G. Gualtieri, A novel method for wind farm layout optimization based on wind turbine selection. Energy Convers. Manag. **193**(April), 106–123 (2019)
129. H. Sun, H. Yang, X. Gao, Investigation into spacing restriction and layout optimization of wind farm with multiple types of wind turbines. Energy **168**(2019), 637–650 (2019)
130. L. Wang, M.E. Cholette, Y. Zhou, J. Yuan, A.C.C. Tan, Y. Gu, Effectiveness of optimized control strategy and different hub height turbines on a real wind farm optimization. Renew. Energy **126**, 819–829 (2018)
131. F. Haces-Fernandez, H. Li, D. Ramirez, Improving wind farm power output through deactivating selected wind turbines. Energy Convers. Manag. **187**(March), 407–422 (2019)
132. F. Toja-Silva, T. Kono, C. Peralta, O. Lopez-Garcia, J. Chen, A review of computational fluid dynamics (CFD) simulations of the wind flow around buildings for urban wind energy exploitation. J. Wind Eng. Ind. Aerodyn. **180**(July), 66–87 (2018)
133. F. Toja-Silva, C. Peralta, O. Lopez-Garcia, J. Navarro, I. Cruz, Roof region dependent wind potential assessment with different RANS turbulence models. J. Wind Eng. Ind. Aerodyn. **142**, 258–271 (2015)
134. I. Abohela, N. Hamza, S. Dudek, Effect of roof shape, wind direction, building height and urban configuration on the energy yield and positioning of roof mounted wind turbines. Renew. Energy **50**, 1106–1118 (2013)
135. A.S. Yang, Y.M. Su, C.Y. Wen, Y.H. Juan, W.S. Wang, C.H. Cheng, Estimation of wind power generation in dense urban area. Appl. Energy **171**, 213–230 (2016)
136. S. Watson, A. Moro, V. Reis, C. Baniotopoulos, S. Barth, G. Bartoli, F. Bauer, E. Boelman, D. Bosse, A. Cherubini, A. Croce, L. Fagiano, M. Fontana, A. Gambier, K. Gkoumas, C. Golightly, M.I. Latour, P. Jamieson, J. Kaldellis, A. Macdonald, J. Murphy, M. Muskulus, F. Petrini, L. Pigolotti, F. Rasmussen, P. Schild, R. Schmehl, N. Stavridou, J. Tande, N. Taylor, T. Telsnig, R. Wiser, Future emerging technologies in the wind power sector: a European perspective. Renew. Sustain. Energy Rev. **113**(June), 109270 (2019)
137. J. Dai, W. Yang, J. Cao, D. Liu, X. Long, Ageing assessment of a wind turbine over time by interpreting wind farm SCADA data. Renew. Energy **116**, 199–208 (2018)
138. A.G. Alexandrov, V.N. Chestnov, V.A. Alexandrov, Identification based control for wind turbine. IFAC-PapersOnLine **50**(1), 2272–2277 (2017)
139. M. Narayana, K.M. Sunderland, G. Putrus, M.F. Conlon, Adaptive linear prediction for optimal control of wind turbines. Renew. Energy **113**, 895–906 (2017)
140. A. Azizi, H. Nourisola, S. Shoja-Majidabad, Fault tolerant control of wind turbines with an adaptive output feedback sliding mode controller. Renew. Energy **135**, 55–65 (2019)
141. M.G. Khalfallah, A.M. Koliub, Effect of dust on the performance of wind turbines. Desalination **209**(1–3 SPEC ISS.), 209–220 (2007)
142. E. Sagol, M. Reggio, A. Ilinca, Issues concerning roughness on wind turbine blades. Renew. Sustain. Energy Rev. **23**, 514–525 (2013)
143. A. González-González, D. Galar, Condition monitoring of wind turbine pitch controller: a maintenance approach. Measurement **123**(January), 80–93 (2017)
144. M.L. Corradini, G. Ippoliti, G. Orlando, An observer-based blade-pitch controller of wind turbines in high wind speeds. Control Eng. Pract. **58**(Feb 2016), 186–192 (2017)

145. J. Lan, R.J. Patton, X. Zhu, Fault-tolerant wind turbine pitch control using adaptive sliding mode estimation. Renew. Energy **116**, 219–231 (2018)
146. V. Irizar, C.S. Andreasen, Hydraulic pitch control system for wind turbines: advanced modeling and verification of an hydraulic accumulator. Simul. Model. Pract. Theory **79**, 1–22 (2017)
147. American Roller Bearing Company, Friction & Frequency Factors (2013)
148. ONYX InSight, Wind Turbine Fire due to Mechanical Failure (2018)
149. R. Brooks, Lessons Learned: common wind turbine bearing failures, costs & solutions, in *4th Wind Operations & Maintenance Canada 2017 Conference*, Toronto (2017)
150. O. Tonks, Q. Wang, The detection of wind turbine shaft misalignment using temperature monitoring. CIRP J. Manuf. Sci. Technol. **17**, 71–79 (2017)
151. H. Polinder, F.F.A. Van Der Pijl, G.J. De Vilder, P.J. Tavner, Comparison of direct-drive and geared generator concepts for wind turbines. IEEE Trans. Energy Convers. **21**(3), 725–733 (2006)
152. C.M.C.G. Fernandes, L. Blazquez, J. Sanesteban, R.C. Martins, J.H.O. Seabra, Energy efficiency tests in a full scale wind turbine gearbox. Tribol. Int. **101**, 375–382 (2016)
153. J.P. Salameh, S. Cauet, E. Etien, A. Sakout, L. Rambault, Gearbox condition monitoring in wind turbines: a review. Mech. Syst. Signal Process. **111**, 251–264 (2018)
154. S. Shanbr, F. Elasha, M. Elforjani, J. Teixeira, Detection of natural crack in wind turbine gearbox. Renew. Energy **118**, 172–179 (2018)
155. U. Bhardwaj, A.P. Teixeira, C.G. Soares, Reliability prediction of bearings of an offshore wind turbine gearbox in *Advances in Renewable Energies Offshore—Proceedings of the 3rd International Conference on Renewable Energies Offshore, RENEW 2018* vol. 141, pp. 779–787 (2019)
156. W. Teng, X. Ding, Y. Zhang, Y. Liu, Z. Ma, Application of cyclic coherence function to bearing fault detection in a wind turbine generator under electromagnetic vibration. Mech. Syst. Signal Process. **87**(June 2016), 279–293 (2017)
157. J. Lloberas, A. Sumper, M. Sanmarti, X. Granados, A review of high temperature superconductors for offshore wind power synchronous generators (2014)
158. W.C. Sant'Ana, C.P. Salomon, G. Lambert-Torres, L.E. Borges da Silva, E.L. Bonaldi, L.E. de Lacerda de Oliveira, J.G.B. da Silva, Early detection of insulation failures on electric generators through online Frequency Response Analysis. Electr. Power Syst. Res. **140**, 337–343 (2016)
159. X. Jin, J. Wenbin, Z. Zhang, L. Guo, X. Yang, System safety analysis of large wind turbines. Renew. Sustain. Energy Rev. **56**, 1293–1307 (2016)
160. M.-Y. Cheng, Y.-F. Wu, Y.-W. Wu, S. Ndure, Fuzzy Bayesian schedule risk network for offshore wind turbine installation. Ocean Eng. **188**(Dec 2018), 106238 (2019)
161. X. Liu, L. Bo, H. Luo, Dynamical measurement system for wind turbine fatigue load. Renew. Energy **86**, 909–921 (2016)
162. Z. Fan, C. Zhu, The optimization and the application for the wind turbine power-wind speed curve. Renew. Energy **140**, 52–61 (2019)
163. The Royal Academy of Engineering, Wind Turbine Power Calculations. Technical report, RWE Npower Ltd. (2009)
164. K.A. Connors, *Chemical Kinetics: The Study of Reaction Rates in Solution* (Wiley-Vch, New York, NY, 1990)
165. W.C. Gardiner Jr. (ed.), *Gas-Phase Combustion Chemistry*, 2nd edn. (Springer, New York, Austin, TX, 2000)
166. T. Turányi, A.S. Tomlin, *Analysis of Kinetic Reaction Mechanisms* (Springer, Berlin, 2014)
167. T. Poinsot, D. Veynante, *Theoretical and Numerical Combustion*, 2nd edn. (Edwards Inc., Philadelphia, USA, 2005)
168. C.K. Law, *Combustion Physics* (Cambridge University Press, NJ, 2010)
169. T. Lieuwen, *Unsteady Combustor Physics* (Cambridge University Press, New York, NY, 2012)
170. R.J. Reed, *North American Combustion Handbook*, vol. 1, 3rd edn. (North American Mfg. Co., Claveland, OH, 1986)

171. R.J. Reed, *North American Combustion Handbook*, vol. 2, 3rd edn. (North American Mfg. Co., Cleveland, OH, 1997)
172. P. Basu, C. Kefa, L. Jestin, *Boilers and Burners: Design and Theory* (Springer, New York, NY, 2000)
173. A.H. Lefebvre, D.R. Ballal, *Gas Turbine Combustion*, 3rd edn. (CRC Press, Boca Raton, 2010)
174. N. Abas, A. Kalair, N. Khan, Review of fossil fuels and future energy technologies. Futures **69**, 31–49 (2015)
175. P. Bórawski, A. Bełdycka-Bórawska, E.J. Szymańska, K.J. Jankowski, B. Dubis, J.W. Dunn, Development of renewable energy sources market and biofuels in The European Union. J. Clean. Prod. **228**, 467–484 (2019)
176. T. Wilberforce, Z. El-Hassan, F.N. Khatib, A. Al Makky, A. Baroutaji, J.G. Carton, A.G. Olabi, Developments of electric cars and fuel cell hydrogen electric cars. Int. J. Hydrog. Energy **42**(40), 25695–25734 (2017)
177. J. Shin, W.S. Hwang, H. Choi, Can hydrogen fuel vehicles be a sustainable alternative on vehicle market?: Comparison of electric and hydrogen fuel cell vehicles. Technol. Forecast. Soc. Change **143**(Jan), 239–248 (2019)
178. M. Wang, R. Dewil, K. Maniatis, J. Wheeldon, T. Tan, J. Baeyens, Y. Fang, Biomass-derived aviation fuels: challenges and perspective. Prog. Energy Combust. Sci. **74**, 31–49 (2019)
179. H. Wei, W. Liu, X. Chen, Q. Yang, J. Li, H. Chen, Renewable bio-jet fuel production for aviation: a review (2019)
180. H.-G. Chen, Y.-H.P. Zhang, New biorefineries and sustainable agriculture: Increased food, biofuels, and ecosystem security. Renew. Sustain. Energy Rev. **47**, 117–132 (2015)
181. C. Zheng, Z. Liu (eds.), *Oxy-Fuel Combustion: Fundamentals, Theory and Practice* (Academic Press, 2018)
182. R. Prieler, P. Bělohradský, B. Mayr, A. Rinner, C. Hochenauer, Validation of turbulence/chemistry interaction models for use in Oxygen enhanced combustion. Energy Procedia **120**, 548–555 (2017)
183. F. Xing, A. Kumar, Y. Huang, S. Chan, C. Ruan, G. Sai, X. Fan, Flameless combustion with liquid fuel: a review focusing on fundamentals and gas turbine application. Appl. Energy **193**, 28–51 (2017)
184. A. Valera-Medina, H. Xiao, M. Owen-Jones, W.I.F. David, P.J. Bowen, Ammonia for power. Prog. Energy Combust. Sci. **69**, 63–102 (2018)
185. M.G. Božo, M.O. Vigueras-Zuniga, M. Buffi, T. Seljak, A. Valera-Medina, Fuel rich ammonia-hydrogen injection for humidified gas turbines. Appl. Energy **251**(Dec 2018) (2019)
186. B. Anderhofstadt, S. Spinler, Factors affecting the purchasing decision and operation of alternative fuel-powered heavy-duty trucks in Germany–A Delphi study. Transport. Res. Part D: Transp. Environ. 73(Nov 2018):87–107, 2019
187. J.M. Beér, N.A. Chigier, *Combustion Aerodynamics* (Robert E. Krieger Publishing Company Inc, London, 1972)
188. R. Borghi, Turbulent combustion modelling (1988)
189. D. Veynante, L. Vervisch, Turbulent combustion modeling. Prog. Energy Combust. Sci. **28**(3), 193–266 (2002)
190. E.D. Gonzalez-Juez, A.R. Kerstein, R. Ranjan, S. Menon, Advances and challenges in modeling high-speed turbulent combustion in propulsion systems. Prog. Energy Combust. Sci. **60**, 26–67 (2017)
191. F.Q. Zhao, H. Hiroyasu, The applications of laser Rayleigh scattering to combustion diagnostics. Prog. Energy Combust. Sci. **19**(6), 447–485 (1993)
192. K. Kohse-Höinghaus, R.S. Barlow, M. Aldén, J. Wolfrum, Combustion at the focus: laser diagnostics and control. Proc. Combust. Inst. **30**(1), 89–123 (2005)
193. R.S. Barlow, Laser diagnostics and their interplay with computations to understand turbulent combustion. Proc. Combust. Inst. **31**I(1), 49–75 (2007)
194. H.A. Michelsen, Probing soot formation, chemical and physical evolution, and oxidation: a review of in situ diagnostic techniques and needs. Proc. Combust. Inst. **36**(1), 717–735 (2017)

195. J. Jedelsky, M. Maly, N.P. del Corral, G. Wigley, L. Janackova, M. Jicha, Air–liquid interactions in a pressure-swirl spray. Int. J. Heat Mass Transf. **121**, 788–804 (2018)
196. Z. Petranović, W. Edelbauer, M. Vujanović, N. Duić, Modelling of spray and combustion processes by using the Eulerian multiphase approach and detailed chemical kinetics. Fuel **191**, 25–35 (2017)
197. A. Urbán, V. Józsa, A. Urbán, V. Józsa, A. Urbán, V. Józsa, Investigation of fuel atomization with density functions. Period. Polytech. Mech. Eng. **62**(1), 33–41 (2018)
198. D. Csemány, V. Józsa, Fuel Evaporation in an atmospheric premixed burner: sensitivity analysis and spray vaporization. Processes **5**(4), 80 (2017)
199. E. Filimonova, A. Bocharov, V. Bityurin, Influence of a non-equilibrium discharge impact on the low temperature combustion stage in the HCCI engine. Fuel **228**, 309–322 (2018)
200. X. Liu, Y. Sage Kokjohn, L.H. Wang, H. Li, M. Yao, A numerical investigation of the combustion kinetics of reactivity controlled compression ignition (RCCI) combustion in an optical engine. Fuel **241**, 753–766 (2019)
201. T. Pachiannan, W. Zhong, S. Rajkumar, Z. He, X. Leng, Q. Wang, A literature review of fuel effects on performance and emission characteristics of low-temperature combustion strategies (2019)
202. National Institute of Standards and Technology, Material Measurement Laboratory (2019). www.nist.gov/mml
203. I. Glassman, R. Yetter, *Combustion*, 4th edn. (Academic Press, Burlington, 2008)
204. S. Tsuboi, S. Miyokawa, M. Matsuda, T. Yokomori, N. Iida, Influence of spark discharge characteristics on ignition and combustion process and the lean operation limit in a spark ignition engine. Appl. Energy **250**(January), 617–632 (2019)
205. P. Glarborg, J.A. Miller, B. Ruscic, S.J. Klippenstein, Modeling nitrogen chemistry in combustion. Prog. Energy Combust. Sci. **67**, 31–68 (2018)
206. I.M. Kennedy, Models of soot formation and oxidation. Prog. Energy Combust. Sci. **23**(2), 95–132 (1997)
207. A.E. Karataş, Ö.L. Gülder, Soot formation in high pressure laminar diffusion flames. Prog. Energy Combust. Sci. **38**(6), 818–845 (2012)
208. Y. Wang, S.H. Chung, Soot formation in laminar counterflow flames. Prog. Energy Combust. Sci. **74**, 152–238 (2019)
209. S.M. Correa, A review of NOx formation under gas-turbine combustion conditions. Combust. Sci. Technol. **87**(1–6), 329–362 (1993)
210. A.G. Gaydon, *The Spectroscopy of Flames*, 2nd edn. (Chapman and Hall Ltd., London, 1974)
211. V. Józsa, A. Kun-balog, Spectroscopic analysis of crude rapeseed oil flame. Fuel Process. Technol. **139**, 61–66 (2015)
212. C.T. Chong, M.-C. Chiong, J.-H. Ng, M. Lim, M.-V. Tran, A. Valera-Medina, W.W.F. Chong, Oxygenated sunflower biodiesel: spectroscopic and emissions quantification under reacting swirl spray conditions. Energy **178**, 804–813 (2019)
213. U. Maas, S.B. Pope, Simplifying chemical kinetics: intrinsic low-dimensional manifolds in composition space. Combust. Flame **88**(3–4), 239–264 (1992)
214. ANSYS Inc. ANSYS Fluent Theory Guide 2019 R2 (2019)
215. S.A. Channiwala, P.P. Parikh, A unified correlation for estimating HHV of solid, liquid and gaseous fuels. Fuel **81**(8), 1051–1063 (2002)
216. Y. Shi, X. Zhang, F. Li, L. Ma, Engineering acid dew temperature: the limitation for flue gas heat recovery. Chin. Sci. Bull. **59**(33), 4418–4425 (2014)
217. A.P. Rossiter, B.P. Jones (eds.), *Energy Management and Efficiency for the Process Industries* (Wiley, 2015)
218. H. Struchtrup, *Thermodyn. Energy Convers.* (Springer, Heidelberg, 2014)
219. B. Kılkış, Ş. Kılkış, New exergy metrics for energy, environment, and economy nexus and optimum design model for nearly-zero exergy airport (nZEXAP) systems. Energy **140**, 1329–1349 (2017)
220. C. Michalakakis, J.M. Cullen, A.G. Hernandez, B. Hallmark, Exergy and network analysis of chemical sites. Sustain. Prod. Consum. 1–19 (2019)

221. Y. Huang, V. Yang, Dynamics and stability of lean-premixed swirl-stabilized combustion. Prog. Energy Combust. Sci. **35**(4), 293–364 (2009)
222. W. Meier, X.R. Duan, P. Weigand, Investigations of swirl flames in a gas turbine model combustor, II. Turbulence-chemistry interactions. Combust. Flame **144**(1–2), 225–236 (2006)
223. S. Taamallah, Z.A. LaBry, S.J. Shanbhogue, M.A.M. Habib, A.F. Ghoniem, Correspondence between "Stable" flame macrostructure and thermo-acoustic instability in premixed swirl-stabilized turbulent combustion. J. Eng. Gas Turbines Power **137**(7), 071505 (2015)
224. R.W. Francisco, A.A.M. Oliveira, Simultaneous measurement of the adiabatic flame velocity and overall activation energy using a flat flame burner and a flame asymptotic model. Exp. Therm. Fluid Sci. **90**(Mar 2017), 174–185 (2018)
225. X. Han, Z. Wang, S. Wang, R. Whiddon, Y. He, Y. Lv, A.A. Konnov, Parametrization of the temperature dependence of laminar burning velocity for methane and ethane flames. Fuel **239**(Nov 2018), 1028–1037 (2019)
226. A.A. Konnov, A. Mohammad, V.R. Kishore, N.I. Kim, C. Prathap, S. Kumar, A comprehensive review of measurements and data analysis of laminar burning velocities for various fuel+air mixtures. Prog. Energy Combust. Sci. **68**, 197–267, (2018)
227. Bioenergy Advice, Composition of wood (2019)
228. S. Van Loo, J. Koppejan (eds.), *The Handbook of Biomass Combustion and Co-firing* (Routledge, London, 2008)
229. E. Fooladgar, P. Tóth, C. Duwig, Characterization of flameless combustion in a model gas turbine combustor using a novel post-processing tool. Combust. Flame **204**, 356–367 (2019)
230. A.A.V. Perpignan, A.G. Rao, D.J.E.M. Roekaerts, Flameless combustion and its potential towards gas turbines (2018)
231. K.I. Khidr, Y.A. Eldrainy, M.M. EL-Kassaby, Towards lower gas turbine emissions: flameless distributed combustion. Renew. Sustain. Energy Rev. **67**, 1237–1266 (2017)
232. R. Amirante, P. De Palma, E. Distaso, P. Tamburrano, Thermodynamic analysis of small-scale externally fired gas turbines and combined cycles using turbo-compound components for energy generation from solid biomass. Energy Convers. Manag. **166**(April), 648–662 (2018)
233. O. Olumayegun, M. Wang, G. Kelsall, Closed-cycle gas turbine for power generation: a state-of-the-art review. Fuel **180**, 694–717 (2016)
234. J. Sachdeva, O. Singh, Thermodynamic analysis of solar powered triple combined Brayton, Rankine and Organic Rankine Cycle for carbon free power. Renew. Energy **139**(2019), 765–780 (2019)
235. C. Bernardoni, M. Binotti, A. Giostri, Techno-economic analysis of closed OTEC cycles for power generation. Renew. Energy **132**, 1018–1033 (2019)
236. C. Breidenich, D. Magraw, A. Rowley, J.W. Rubin, The Kyoto protocol to the united nations framework convention on climate change. Am. J. Int. Law **92**(2), 315 (1998)
237. M. Yang, H. Zhang, Z. Meng, Y. Qin, Experimental study on R1234yf/R134a mixture (R513A) as R134a replacement in a domestic refrigerator. Appl. Therm. Eng. **146**(Sept 2018), 540–547 (2019)
238. V. Pethurajan, S. Sivan, G.C. Joy, Issues, comparisons, turbine selections and applications—an overview in Organic Rankine Cycle. Energy Convers. Manag. **166**(March), 474–488 (2018)
239. K.A. Abrosimov, A. Baccioli, A. Bischi, Techno-economic analysis of combined inverted Brayton-Organic Rankine Cycle for high-temperature waste heat recovery. Energy Convers. Manag.: X **3**(June), 100014 (2019)
240. Q. Sun, Y. Wang, Z. Cheng, J. Wang, P. Zhao, Y. Dai, Thermodynamic optimization of a double-pressure Organic Rankine Cycle driven by geothermal heat source. Energy Procedia **129**, 591–598 (2017)
241. W.R. Huster, D. Bongartz, A. Mitsos, Deterministic global optimization of the design of a geothermal Organic Rankine Cycle. Energy Procedia **129**, 50–57 (2017)
242. S. Van Erdeweghe, J. Van Bael, B. Laenen, W. D'haeseleer, Design and off-design optimization procedure for low-temperature geothermal Organic Rankine Cycles. Appl. Energy **242**(Feb), 716–731 (2019)

243. M. Ahmad, M. Schatz, M.V. Casey, Experimental investigation of droplet size influence on low pressure steam turbine blade erosion. Wear **303**(1–2), 83–86 (2013)

244. G. Györke, U.K. Deiters, A. Groniewsky, I. Lassu, A.R. Imre, Novel classification of pure working fluids for Organic Rankine Cycle. Energy **145**, 288–300 (2018)

245. A. Groniewsky, G. Györke, A.R. Imre, Description of wet-to-dry transition in model ORC working fluids. Appl. Therm. Eng. **125**, 963–971 (2017)

246. A. Groniewsky, A.R. Imre, Prediction of the ORC working fluid's temperature-entropy saturation boundary using redlich-Kwong equation of state. Entropy **20**(2), 1–8 (2018)

247. A.R. Imre, R. Kustán, A. Groniewsky, Thermodynamic selection of the optimal working fluid for Organic Rankine Cycles. Energies **12**(10), 1–15 (2019)

248. Y. Zhao, G. Liu, L. Li, Q. Yang, B. Tang, Y. Liu, Expansion devices for Organic Rankine Cycle (ORC) using in low temperature heat recovery: a review. Energy Convers. Manag. **199**(August), 111944 (2019)

249. M. Saghafifar, A. Omar, K. Mohammadi, A. Alashkar, M. Gadalla, A review of unconventional bottoming cycles for waste heat recovery: Part I—analysis, design, and optimization. Energy Convers. Manag. 1–59 (2018)

250. A. Omar, M. Saghafifar, K. Mohammadi, A. Alashkar, M. Gadalla, A review of unconventional bottoming cycles for waste heat recovery: Part II—applications. Energy Convers. Manag. **180**(Sept 2018), 559–583 (2019)

251. W.B Nader, C. Mansour, C. Dumand, M. Nemer, Brayton cycles as waste heat recovery systems on series hybrid electric vehicles. Energy Convers. Manag. **168**(Feb), 200–214 (2018)

252. X. Li, H. Tian, G. Shu, M. Zhao, C.N. Markides, H. Chen, Potential of carbon dioxide transcritical power cycle waste-heat recovery systems for heavy-duty truck engines. Appl. Energy **250**(May), 1581–1599 (2019)

253. P. Liu, G. Shu, H. Tian, How to approach optimal practical Organic Rankine Cycle (OP-ORC) by configuration modification for diesel engine waste heat recovery. Energy **174**, 543–552 (2019)

254. C. Falter, R. Pitz-Paal, Energy analysis of solar thermochemical fuel production pathway with a focus on waste heat recuperation and vacuum generation. Sol. Energy **176**(September), 230–240 (2018)

255. M. Awais, A.A. Bhuiyan, Recent advancements in impedance of fouling resistance and particulate depositions in heat exchangers. Int. J. Heat Mass Transf. **141**, 580–603 (2019)

256. E. Wallhäußer, M.A. Hussein, T. Becker, Detection methods of fouling in heat exchangers in the food industry. Food Control **27**(1), 1–10 (2012)

257. M.J. Li, S.Z. Tang, F.l. Wang, Q.X. Zhao, W.Q. Tao, Gas-side fouling, erosion and corrosion of heat exchangers for middle/low temperature waste heat utilization: a review on simulation and experiment. Appl. Therm. Eng. **126**, 737–761 (2017)

258. M. Trafczynski, M. Markowski, K. Urbaniec, Energy saving potential of a simple control strategy for heat exchanger network operation under fouling conditions. Renew. Sustain. Energy Rev. **111**(May), 355–364 (2019)

259. S.K. Singh, M. Mishra, P.K. Jha, Nonuniformities in compact heat exchangers—scope for better energy utilization: a review. Renew. Sustain. Energy Rev. **40**, 583–596 (2014)

260. S. Wang, Y. Xinquan, C. Liang, Y. Zhang, Enhanced condensation heat transfer in air-conditioner heat exchanger using superhydrophobic foils. Appl. Therm. Eng. **137**(April), 758–766 (2018)

261. L. Herraiz, D. Hogg, J. Cooper, M. Lucquiaud, Reducing the water usage of post-combustion capture systems: The role of water condensation/evaporation in rotary regenerative gas/gas heat exchangers. Appl. Energy **239**(July 2018), 434–453 (2019)

262. F. Chemat, N. Rombaut, A. Meullemiestre, M. Turk, S. Perino, A.S. Fabiano-Tixier, M. Abert-Vian, Review of green food processing techniques. Preservation, transformation, and extraction. Innov. Food Sci. Emerg. Technol. **41**(May), 357–377 (2017)

263. M.V.D. Bonis, G. Ruocco, Heat and mass transfer modeling during continuous flow processing of fluid food by direct steam injection. Int. Commun. Heat Mass Transf. **37**(3), 239–244 (2010)

264. L. Chen, Y.L. Liu, J.L. Deng, Removal of phthalic acid esters from sea buckthorn (Hippophae rhamnoides L.) pulp oil by steam distillation and molecular distillation. Food Chem. **294**(May), 572–577 (2019)

265. X. Meng, Z. Wen, Y. Qian, Y. Hongbing, Evaluation of cleaner production technology integration for the Chinese herbal medicine industry using carbon flow analysis. J. Clean. Prod. **163**, 49–57 (2017)

266. F. Memarzadeh, Adding amines to steam for humidification. J. Chem. Health Saf. **21**(4), 5–17 (2014)

267. V. Gorobets, Y. Bohdan, V. Trokhaniak, I. Antypov, Investigations of heat transfer and hydrodynamics in heat exchangers with compact arrangements of tubes. Appl. Therm. Eng. **151**(Dec 2018), 46–54 (2019)

268. T. Muszynski, The influence of microjet array area ratio on heat transfer in the compact heat exchanger. Exp. Therm. Fluid Sci. **99**(July), 336–343 (2018)

269. M. Awais, A.A. Bhuiyan, Heat and mass transfer for compact heat exchanger (CHXs) design: a state-of-the-art review. Int. J. Heat Mass Transf. **127**, 359–380 (2018)

270. H. Mroue, J.B. Ramos, L.C. Wrobel, H. Jouhara, Experimental and numerical investigation of an air-to-water heat pipe-based heat exchanger. Appl. Therm. Eng. **78**, 339–350 (2015)

271. J. Choi, M. Jeong, J. Yoo, M. Seo, A new CPU cooler design based on an active cooling heatsink combined with heat pipes. Appl. Therm. Eng. **44**, 50–56 (2012)

272. K.S. Kim, M.H. Won, J.W. Kim, B.J. Back, Heat pipe cooling technology for desktop PC CPU. Appl. Therm. Eng. **23**(9 SPEC.), 1137–1144 (2003)

273. G. Zhou, J. Li, L. Lv, An ultra-thin miniature loop heat pipe cooler for mobile electronics. Appl. Therm. Eng. **109**, 514–523 (2016)

274. H. Shabgard, M.J. Allen, N. Sharifi, S.P. Benn, A. Faghri, T.L. Bergman, Heat pipe heat exchangers and heat sinks: opportunities, challenges, applications, analysis, and state of the art. Int. J. Heat Mass Transf. **89**, 138–158 (2015)

275. C. Wang, J. Chen, S. Qiu, W. Tian, D. Zhang, G.H. Su, Performance analysis of heat pipe radiator unit for space nuclear power reactor. Ann. Nucl. Energy **103**, 74–84 (2017)

276. L. Ge, L. Huaqi, S. Jianqiang, Reliability and loading-following studies of a heat pipe cooled, AMTEC conversion space reactor power system. Ann. Nucl. Energy **130**, 82–92 (2019)

277. H.U. Oh, S. Shin, C.W. Baek, Thermal control of spaceborne image sensor using heat pipe cooling system. Aerosp. Sci. Technol. **29**(1), 394–402 (2013)

278. B. Zohuri, *Heat Pipe Design and Technology*, 2nd edn. (Springer International Publishing, 2016)

279. M. Liao, Z. He, C. Jiang, X. Fan, Y. Li, F. Qi, A three-dimensional model for thermoelectric generator and the influence of Peltier effect on the performance and heat transfer. Appl. Therm. Eng. **133**(January), 493–500 (2018)

280. R. Merienne, J. Lynn, E. McSweeney, S.M. O'Shaughnessy, Thermal cycling of thermoelectric generators: the effect of heating rate. Appl. Energy **237**(Nov 2018), 671–681 (2019)

281. P. Wang, J.E. Li, B.L. Wang, T. Shimada, H. Hirakata, C. Zhang, Lifetime prediction of thermoelectric devices under thermal cycling. J. Power Sources **437**(June), 226861 (2019)

282. J.W. Stevens, Optimal design of small ΔT thermoelectric generation systems. Energy Convers. Manag. **42**(6), 709–720 (2001)

283. P. Wang, B.L. Wang, J.E. Li, Temperature and performance modeling of thermoelectric generators. Int. J. Heat Mass Transf. **143**, 118509 (2019)

284. A. Allouhi, Advances on solar thermal cogeneration processes based on thermoelectric devices: a review. Sol Energy Mater. Sol. Cells **200**(May), 109954 (2019)

285. A. Singha, Optimized Peltier cooling via an array of quantum dots with stair-like ground-state energy configuration. Phys. Lett. Section A: Gen. At. Solid State Phys. **382**(41), 3026–3030 (2018)

286. H.H. Saber, S.A. AlShehri, W. Maref, Performance optimization of cascaded and non-cascaded thermoelectric devices for cooling computer chips. Energy Convers. Manag. **191**(April), 174–192 (2019)

287. C. Lundgaard, O. Sigmund, Design of segmented thermoelectric Peltier coolers by topology optimization. Appl. Energy **239**(July 2018), 1003–1013 (2019)
288. Y. Lyu, A.R.M. Siddique, S.H. Majid, M. Biglarbegian, S.A. Gadsden, S. Mahmud, Electric vehicle battery thermal management system with thermoelectric cooling. Energy Rep. **5**, 822–827 (2019)
289. L. van Dommelen, *Quantum Mechanics for Engineers* (2018)
290. G. Lebon, D. Jou, *Understanding Non-equilibrium Thermodynamics* (Springer, Berlin, 2008)

Chapter 4
Thermal Processes in Vacuum

1 Vacuum Applications

The most straightforward application of heat transfer in the absence of convection is space exploration which is now turning into industry with the involvement of private companies. Nevertheless, thermal aspects of vacuum applications are also common in both research and industrial processes. These are, e.g., electron microscopes [1–3], scanning thermal microscopy for local heat conduction measurement [4, 5], desalination [6–8], waste treatment [9, 10], solar collector manufacturing [11–13], and vacuum heat treatment [14, 15]. Cryogenics will not be discussed in this book due to the numerous engineering challenges. Nevertheless, the presented methods in this book provide a solid theoretical background; this is the easier part. See Ref. [16] for rich practical details on the topic which are the key to design such a device or facility.

2 Thermal Radiation

Thermal radiation is the dominant mode of heat propagation in vacuum. While the other ways (conduction and convection) require a medium, it does not hold for the radiative heat transfer as the propagation is occurring in the form of electromagnetic waves. In general, the emitted heat flux depends on the direction which is neglected now for simplicity which is justifiable in thermal engineering applications. Thermal radiation can be characterized using the total power, $\dot{Q}$, and wavelength, λ_w. Using these quantities, the intensity is calculated as:

$$I = \frac{\mathrm{d}\dot{Q}}{\mathrm{d}A\mathrm{d}\lambda_w}, \tag{1}$$

where A is the emitting surface. Thus the intensity is the total radiated power per unit wavelength times unit area. When that radiation with a given $\dot{Q}$ reaches a body,

© Springer Nature Switzerland AG 2020
V. Józsa and R. Kovács, *Solving Problems in Thermal Engineering*, Power Systems,
https://doi.org/10.1007/978-3-030-33475-8_4

they interact. This also depends on the material properties which could be sensitive to temperature conditions. Some part of the radiation is

- absorbed in the body, it transforms into a space-dependent heat source, and its amount is denoted by $\dot{Q}_a$;
- reflected back to the environment, denoted by $\dot{Q}_r$;
- transmitted and does not affect the body thermally, denoted by $\dot{Q}_t$.

Introducing the notation for the normalized values such as

$$\alpha = \frac{\dot{Q}_a}{\dot{Q}}, \quad \rho = \frac{\dot{Q}_r}{\dot{Q}}, \quad \tau = \frac{\dot{Q}_t}{\dot{Q}}, \tag{2}$$

one can describe the interaction using these quantities, called absorptivity, reflectivity, and transmissivity, respectively. Since the energy balance must be satisfied, that is:

$$\dot{Q} = \dot{Q}_a + \dot{Q}_r + \dot{Q}_t, \tag{3}$$

therefore $\alpha + \rho + \tau = 1$ holds. In the most general case, these coefficients depend on the wavelength and temperature. However, that would be too complicated to implement into a practical application and measure all the required parameters. For the sake of simplicity, 'special' bodies are used to model the real ones, for instance,

- white body: $\rho = 1$, i.e., reflects any radiation,
- transparent body: $\tau = 1$, i.e., transmits all the incident radiation,
- black body: $\alpha = 1$, i.e., absorbs the whole incident radiation,
- grey body: $0 < \alpha < 1$, which is the mix of the others.

In the above cases, the following conditions are emphasized:

- Regarding these 'special' bodies, their behavior is independent of the wavelength.
- The Kirchhoff's law defines the emission factor (or emissivity) as $\varepsilon = \alpha$. Thus the white and transparent bodies produce no emission. Note that this law holds only in thermal equilibrium.

The general one is the color body which is characterized by wavelength-dependent thermal radiation properties. Considering a black body, which is characterized by the Planck's law, i.e., the intensity is given as a function of the temperature T and the wavelength λ_w, $I = I(\lambda_w, T)$. According to the Stefan–Boltzmann law, integrating the function $I(\lambda_w, T)$ along constant temperature yields

$$\dot{q} = \int\limits_0^\infty I(\lambda_w, T)\mathrm{d}\lambda_w = \sigma_0 T^4, \tag{4}$$

in which σ_0 is called the Stefan–Boltzmann constant, and (4) is valid only for diffuse black bodies in thermal equilibrium, and $\dot{q}$ represents only the emitted heat flux without any interaction. When another body is present, their radiative thermal interaction is calculated as:

$$\dot{Q}_{1\to2} = A_1\varepsilon_1\varepsilon_2\sigma_0\varphi_{1\to2}(T_1^4 - T_2^4), \tag{5}$$

where the index $1 \to 2$ assumes that the first body is the warmer one, and φ is the so-called view factor. If the result will be negative, then the second body was warmer, and the way of heat transfer is the opposite as it was assumed. This is not a problem during calculations if the analysis is consequent, i.e., the same heat transfer direction is used for the calculation of the heat balance of the other body. Errors always originate from inconsistent notations. Equation (5) expresses the geometric setting, therefore, it is generally usable. Here, the view factor satisfies the relation $\varphi_{1\to2}A_1 = \varphi_{2\to1}A_2$. The emission of a black body radiator as a function of the wavelength peaks at different wavelength values, λ_{max}, depending on its temperature. The relation between these quantities is expressed by Wien's displacement law, which is the direct consequence of the Planck's law

$$\lambda_{max} \cdot T = 2898 \, \mu m \cdot K. \tag{6}$$

For more details on the basics of thermal radiation with examples, see Chap. 8 of Ref. [17]. It also contains few topics which are not discussed here, e.g., gas radiation, and the derivation of thermal resistances for various applications. Practical data on radiation of a wide range of materials is available in Ref. [18].

3 Thermal Control in Vacuum

In the open atmosphere, natural heating/cooling is always present since the air at ambient temperature heats/cools all bodies with lower/higher temperature. Therefore, our devices have to be designed and tested in a narrow temperature range, compared to, e.g., space applications where often $0\,°C$ temperature is desired in the nearly $0\,K$ environment. As vacuum processes on Earth need a reliable pump and special sealing solutions to avoid the intrusion of air, careful thermal design is required to end up with an economically competitive device.

Thermal control is necessary, when the temperature of a device or its component does not meet the desired range when the natural way of heat propagation is provided. If the ambient is cooler, the generated heat should be retained by using either shielding or insulation or both; this is discussed in Sect. 3.1. If there is no inner heat generation or higher temperature present at the beginning of the process, the insulating solutions will provide only a more homogeneous temperature distribution inside which may be also required in certain applications. Section 3.2 discusses the opposite situation: the heat should be released efficiently. The methods described in the two subsections can be used in both space and industrial environment.

3.1 Retaining the Heat: Shielding and Insulation

When thermal shielding is applied, it should be kept in mind that if the shielding provides too good insulation, the device might easily overheat if there is inner heat generation. There are two ways to increase the thermal resistance between a body and its ambient. The first one is using thermal insulation where the body is wrapped with a material that has a low thermal conductivity which is usually in the range of 0.1 W/(m·K). The other way is the use of a few layers of thin foils do reduce the emitted heat by thermal radiation. If desired, the two methods also can be combined.

In both Earth and space, thermal shields often offer the same effect as insulation but in a significantly thinner layer [19]. It is because thermal conduction scales with T while it is T^4 for thermal radiation. Since a gap should be maintained between the layers of the shield and the body, its design may be challenging.

A cross-section of a vacuum tube solar collector, shown in Fig. 1, is an excellent example for retaining the heat. Hence, the methods are discussed through the analysis of this structure since it bears all the important details.

The surface temperature of the Sun is 5500 °C which far exceeds the temperature of any part of the tube, hence, the absorbed heat by the tube can be considered as a heat flux. The majority of the visible light passes the glass envelope and reaches

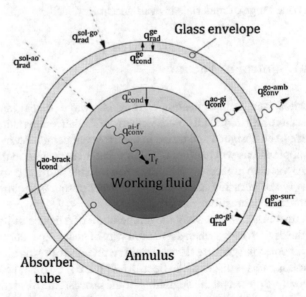

Fig. 1 Heat transfer in a vacuum tube of a solar collector [20]. With permission. Abbreviations: rad: radiation, sol: solar radiation, cond: conduction, conv: convection, g/ge: glass envelope, a: absorber tube, xi/xo: inner/outer surface of x, f: fluid, brack: bracket (not shown)

the wall of the absorber tube, which heats it. There is no heat generation inside by another process, hence, the heat generated by the incident solar flux is used to heat the working fluid via the absorber tube. For efficient operation, the heat propagation from the absorber tube to the ambient has to be minimized. This is solved as follows. The absorptivity of the tube is close to unity in the range of the solar spectrum, however, the emissivity is below 0.1 in the infrared range [21] where the body emits the maximum heat via radiation. Equation (6), Wien's displacement law, answers this phenomenon with different emissivity and absorptivity values. Since the absorbed and emitted radiation is characterized by different wavelength regions with minimal overlap, the wavelength-dependent absorptivity and emissivity allow this operation which is the key to high efficiency. Hence, the surface of the absorber tube is a color body from the thermal radiation point of view. Even though the tube is evacuated, heat conduction might be present between the glass envelope, and the absorber tube as a small amount of gas remains in the tube during its manufacturing.

When the geometry is complex, and there is no room to fix either an insulator or a shield, there are two solutions. The first one is wrapping the whole body with a thermal shield or an insulator with a shiny coating that is practically the combination of shielding and insulation. The second one, which is typically used in cavities, is filling the hole with a non-conductive foam. Since thermal insulators usually good electric insulators as well, this solution keeps the electric circuits safe. However, maintenance after this step is impossible.

A recent approach to store heat in low weight is the use of heat pipes and phase change materials in combination [22]. Since heat pipes are the thermal diodes, they can be used both for retaining or releasing the heat. An example for the latter application is the cooling system of microelectric components, briefly discussed in Sect. 5.3 of Chap. 3.

3.2 Releasing the Heat: Fins and Special Coatings

If there is an excessive heat generation inside of a device, the first thing to consider is its use for, e.g., electricity generation [23]. If there is no use of heat and its rejection is the desired way, there are various solutions available. The most obvious one is using a surface which reflects the majority of the incoming heat while it has high emissivity in the operating wavelength range. This design is practically the inverse of the absorber tube of the solar collector, shown in Fig. 1. Nevertheless, this solution works only when there is a significant temperature difference between the bodies which interact via radiation. In practice, the common materials have roughly equal absorptivity and emissivity over the whole wavelength range of interest in engineering applications. For a rich emissivity data, see Ref. [18].

The use of fins for heat rejection via thermal radiation emerged at the dawn of the space age. An excellent overview of the optimization of fins for such applications can be found in Ref. [24]. Since the price of launching 1 kg mass to space is in the range of 20,000 USD [22], designing a fin only for increasing the heat rejection must have a

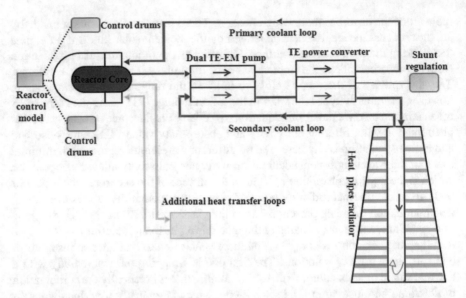

Fig. 2 Schematic of a nuclear reactor used for power generation in space applications [28]. With permission

very good reason. Instead, this is the point when the authors highly encourage the use of numerical software. With few degrees of freedom in the geometry along with their given minimum and maximum size ranges, a genetic algorithm-based optimization tool will surely find the best solution to have the maximum heat rejection while having a low mass. Few examples for this methodology can be found in Refs. [25–27].

A need for heat rejection in space is usually connected to thermoelectric generators, shown in Fig. 2. The heat of the operating reactor is converted into electric power in thermoelectric converter units which are exploiting the Seebeck effect. More details on them can be found in Sect. 5.4 of Chap. 3. These units have no moving parts and require temperature difference on their sides. Hence, the other side requires cooling which are connected to fins designed to reject heat via thermal radiation. The connection of distant objects is usually solved by heat pipes.

4   Thermal Balance in Orbit

The present section highlights the thermal balance in Low Earth Orbit (LEO) since this is the orbit which is affordable for amateur satellite designers [29]. Nevertheless, the presented logic can be used for other space missions as well as a starting point. The communication system of a satellite is the most important part of the device, hence, principally electrical engineers design small-sized spacecrafts. However, the failure rate of amateur satellites is significantly higher than that of professional satellites [30]. A potential source of loss is the lack of thermal analysis and testing.

Before starting the design of any spacecraft, it is advised to calculate its lifetime. Since the space debris is a continuously and rapidly increasing threat to all space missions [31–33], it is highly recommended to pick the lowest applicable orbit which allows the completion of the mission. If the electrical circuits of the satellite are built from commercial components instead of ones designed for space use, be within few years lifetime. Since it is affected by the space weather, i.e., principally by the solar flares emitted by the Sun, any calculations will provide only a rough estimation. Nevertheless, the lifetime increase exponentially with the altitude.

The affordable space exploration started in 1999 with the introduction of CubeSats [34]. Besides the low cost, their development time is also low; there are vendors who sell components to allow more focus on the mission objective and the development of the related parts. Since the mass of CubeSats is limited to 1.33 kg, and the average specific heat capacity of the satellite is below 1 kJ/(kg·K), the temperature variation in orbit is significant. As the satellite is isolated in space, the heat transfer between the satellite and its environment is via thermal radiation. These are the followings [35].

- **Solar flux**. Its value depends on the Sun cycle, and the distance from the Sun. Its average value is 1367 W/m^2 at Earth's orbit which is good for preliminary analysis. Due to the elliptical orbit, the low and high average values are 1322 W/m^2 and 1414 W/m^2, respectively.
- **Radiation to space**. The dark background temperature is 2.7 K. This is always present and is the dominant reason of heat loss.
- **Radiation between the satellite and Earth**. As for Earth, its mean temperature can be assumed as 255 K. Sometimes it is calculated as an incoming heat flux to the satellite, however, it varies with the orbit and hence needs corrections. If a fixed temperature is used, the equations to solve will be simpler to handle; and only the view factor needs to be calculated by basic geometric equations.
- **Albedo**. This is the diffuse reflection of the sunlight from the surface of Earth. Majority of the albedo is originated from the snowy and icy poles and deserts. Since this is depending on the weather, its accurate modeling is challenging, and prediction is even harder. However, an average value can be estimated based on the available Earth observation data.

5   Example: Thermal Evaluation of a PocketQube Class Satellite

This subsection shortly summarizes our work in the thermal behavior evaluation of the SMOG-1 PocketQube satellite [36]. The spacecraft is scheduled to launch by the end of 2019/early 2020. Figure 3 shows the variant prepared for qualification tests. The satellite name comes from its main mission goal: monitoring the Earth to provide data on electrosmog in the frequency range of digital TV stations (430–860 MHz).

Fig. 3 The assembled
satellite model for
qualification tests [36]. With
permission

Even though the standard for this satellite class has been removed from the internet, it is worth to present the thermal challenges of a $5 \times 5 \times 5$ cm size since its heat capacity is even lower than that of CubeSats, hence, it is more exposed to temperature oscillations in orbit.

Since the customers principally pay for the mass of the spacecraft, and the electronic components show a dramatic decrease in size over time, such a tiny platform can host a wide variety of experiments and is expected to return to the space industry in an updated form. The solar panels on the small surfaces cannot provide enough energy to power the communication systems to efficiently transfer large images to the ground. However, compressed text data in the range of 0.5 MB/orbit can be safely downloaded by using a single ground station. Nevertheless, it depends on the orbit. An extended antenna system may allow more data to be downloaded.

In such a small size, the first design principle is that there is no room for a separate framework which host the printed circuit boards (PCB). Therefore, the PCBs serve as both load carriers and thermal shields besides hosting the electric components. If commercial products are used, they typically can withstand $-40-+80\,°C$. In the case of SMOG-1, a Li-ion battery is used for energy storage with an allowed temperature range of $-10-+60\,°C$. Since it was not possible to put a redundant battery due to volume restrictions, the system was designed to being able to operate on a single solar cell as well—hence, in the worst case, it will not be functional in Earth's shadow. As for the other electric components, every system was duplicated to tolerate a single point failure. For this reason, solar panels are working individually on each side. The use of commercial electric components is not a straightforward choice and should be avoided if there is low time for testing. They are only accepted as final parts on the spacecraft if extended rigorous tests approve them.

Since repairing in space is not an option, both finite element modeling (FEM) and thermal network (TN) analysis were used to estimate the behavior. Since the former required roughly five magnitude more computational time, it was used only

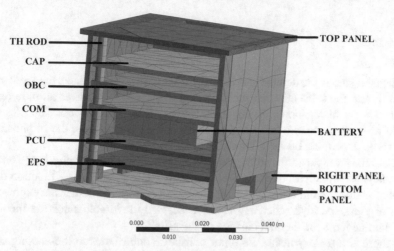

Fig. 4 Simplified structure of the satellite, prepared for FEM analysis [36]. With permission

for checking. All the thermal design was principally performed by using TN. The below subsections detail the calculations of various parts of the model. It was built in Matlab Simulink, however, any other platform can be used for building a similar model; no specific toolboxes/built-in functions were used. The simplified structure of the satellite is shown in Fig. 4, featuring the mesh used for FEM analysis. All the named panels were considered as a single node in the TN analysis, except for the threaded rod which was considered as thermal resistances between the corresponding parts. Hence, 12 units: six side panels, five inner panels, and the battery were analyzed as concentrated parameter subsystems.

5.1   *Thermal Boundary Conditions*

The most important property of a satellite from a thermal analysis point of view is its movement. The orientation determines the thermal radiation received by each surface. When there is no thrust control,[1] it will be a random variable with a minimum, maximum, and expected value.

If a perfect cube is assumed with a side length of unity, and the light is coming from one direction, the minimum and maximum affected surfaces are one and $\sqrt{3}$ square units, respectively. The expected value can be derived by using an integral mean calculation. In this case, it is enough to calculate the orientation from zero to $\pi/4$ by two angles due to symmetry:

---

[1] The sunlight also acts as a force on the surfaces. The most spectacular example for the use of this phenomenon is the revival of the Kepler Space Telescope [37]. It featured six reaction wheels from which the failure of the fourth one anticipated that the mission is over. However, the photon pressure of the sunlight turned to be appropriate for a highly limited but functional 3D motion control.

$$\frac{16}{\pi^2} \int_0^{\pi/4} \int_0^{\pi/4} [\sin(\phi) + \cos(\phi)] \cos(\theta) + \sin(\theta) \mathrm{d}\phi \mathrm{d}\theta = 1.519. \qquad (7)$$

The next step is the determination of the orbit which incorporates the view factor of Earth and the ratio between the dark and lit phases. The lower altitude results in larger view factor, however, the dark phase will last longer, too. The presented results are respective to a 550 km Sun Synchronous Orbit where the lit phase is at least 64%, depending on the orbit of Earth around the Sun.

As for the solar panes, they turn a portion of solar flux into electricity, based on their efficiency. Nevertheless, it is important to know the reflectivity as a function of the angle of incident radiation that also affects the efficiency. From thermal engineering point of view, the high efficiency is not necessarily favorable since the incoming solar flux governs the thermal balance of the satellite.

If there is a battery inside, its inner resistance contributes to the heating of the battery. When the temperature significantly drops, this value notably increases below a certain temperature, meaning a kind of self-control. However, this state should be avoided when the purpose of the battery use is increasing its own temperature. The heating power of the battery used in SMOG-1 produces roughly 10 mW, uniformly distributed in the cells.

If there is no other device on board with high power consumption, i.e., which produces notable amount of heat during its operation, the power amplifier of the radio transmitter will generate the most heat. Therefore, the location of this part should be chosen wisely. In our case, it was placed on the other side of the COM PCB which hosts the battery. This was the best strategy to keep the battery temperature the highest which was the most notable thermal design constrain. More information on the thermal balance of the battery will be discussed in Sect. 5.3.

As for the radiative heat transfer between the ambient and the satellite, appropriate corrections were made to the incoming fluxes to prepare them for employing in both the TN and the FEM analyses which meant averaging and surface projection. The corresponding values and sources are shown in Table 1. The comparison of the two calculation methods will be discussed in Sect. 5.5.

Table 1 Adjusted thermal radiation boundary conditions [36]. With permission

Part	Heat flux source	Magnitude (W/m^2)
Top panel	Incident solar	357
Top solar cell	Incident solar	282
Bottom panel	Albedo + IR	101
Bottom solar cell	Albedo + IR	74
Middle side panels	All three	474
Middle side solar cells	All three	396

5.2 The TN Model

The basics of the TN model were discussed in Sect. 3.2 of Chap. 3. Currently, the key parts of the simulation, including inner connections, are summarized below to give an overview of the required data.

- **Geometry**. The surfaces receive the radiative heat fluxes, and their products are heat flows. In addition, when there is conduction between two surfaces, and the conductor is not occupying the whole surface, the effective area should be estimated to calculate the thermal resistance to be used. The view factors and volumes of each part were also calculated by using geometry data to have the total heat capacities.
- **Material and surface properties**. The thermal conductivity of the FR-4 panels was considered as anisotropic, while this property of other other components were assumed as isotropic. Additional material properties were densities and specific heat capacities. For thermal radiation, absorptivity and emissivity values were estimated based on literature data.
- **Thermal connections**. The following thermal connections were considered. Radiation to space, inner (volumetric) heat generation by electric components, inner thermal radiation, considering the inner view factors, and conduction between the subcomponents. Heat transfer from the ambient: solar flux, albedo, and radiation of Earth.
- **Rotation**. The angular velocity was assumed as one rotation per minute, following basic kinetic calculations of the launch pod. Its direction influences the thermal balance which was the purpose of Eq. (7). As for the magnitude, the solar cells are estimated to work efficiently up to 100 RPM, where the maximum power point tracking algorithm will start losing its full functionality. Note that this loss means an increase in incoming heat flux since more of the solar radiation and albedo will turn into heat. The rotation was added to four sides during the TN modeling while FEM analysis uses only a mean value to end up with reasonable simulation time. It feels that 1 RPM is slow, however, the temperature of an affected side varies less than one °C. Hence, the thermal time scales here are significantly larger than the motion time scales.
- **Transient state**. Due to the orbit, steady-state operation does not exists. Therefore, energy balance equations are calculated for all subcomponents dynamically. Assuming a constant heat capacity, $c \cdot m_c$, it will lead to a first-order inhomogeneous ordinary differential equation for the temperature as:

$$\frac{dT}{dt} = \frac{1}{c \cdot m_c} \cdot \sum \dot{Q}. \tag{8}$$

To solve Eq. (8), the fourth-order Runge–Kutta algorithm was used. This is a standard starting point in engineering problems. If there are discrepancies or unexpected behavior present when the algorithm is checked, only then should one test other methods or when the runtime becomes excessive. For a single node analysis of small satellites, the Runge–Kutta method was considered as a reference by Ahn et al. [38].

5.3   Battery

In theory, a high number of thermal shield layers will reduce the heat transfer via radiation between the battery and the neighboring parts to practically zero. However, overheating should also be avoided, hence, a careful solution is required.

In practice, the battery has two wires which connect it to the energy bus. Hence, heat transfer via thermal conduction will affect its temperature, even though the radiative losses approach zero. Since copper is a good thermal conductor, too, this loss will represent the minimum value. Again, in theory, very thin wires would mean the lowest possible thermal losses, however, the structure has to survive the launch procedure. Consequently, the governing heat transfer way between the battery and the panel of the power control unit will be conduction. Knowing this fact, few layers of shielding are appropriate for minimizing the thermal losses. Nevertheless, if there is room for other temperature control methods, they can be used efficiently [39, 40].

Note that all the thermal protection principally serve as a thermal resistance between the ambient and the satellite. The higher the resistance, the lower the temperature fluctuation over an orbit. Since there is inner heat generation, shiny cover surfaces are better since the majority of the materials have equal emissivity and absorptivity. However, a surface finish with high absorptivity in the visible spectrum and low emissivity in the infrared spectrum is the best to maintain high inner temperature.

The effect of the solar flux on the battery temperature is shown in Fig. 5. even though there is 7% variation between the minimum and the maximum solar flux, the battery absolute temperature will vary only 1.5% since the radiative heat transfer scales with T^4. In addition, it can be concluded that the current conditions are appropriate for the battery; it will operate in the allowed temperature range.

5.4   Overall Thermal Balance

After five orbits, the thermal balance of the satellite became final in the present case, i.e., the initial conditions do not affect the results anymore. As for the orientation,

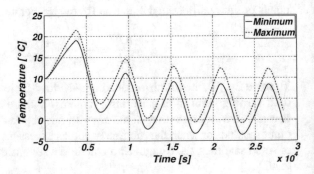

Fig. 5 Temperature variation of the battery at minimum and maximum solar fluxes [36]. With permission

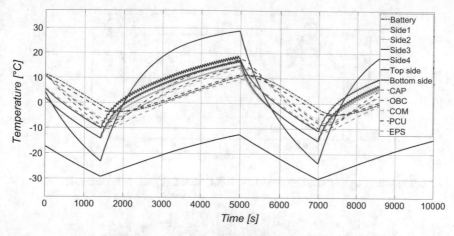

Fig. 6 Temperature variation of all PCBs in orbit at 1322 W/m² solar flux

five panels are considered as lit (four of them are rotating) while one is not lit. The temperature variation is shown in Fig. 6. Since the minimal solar flux was set, and the lowest temperature is 10 °C above the minimum rated temperature of the commercial electric components, it can be concluded that the thermal design at this orbit is appropriate.

The effect of rotation causes less than one °C temperature variation on the sides, shown in Fig. 6. The magenta line represents the battery; it has the lowest temperature variation as it was desired. The claret curve at the bottom represents the bottom side while the blue curve shows the temperature variation of the top side. These sides were not rotated. The high temperature variation of the top side is due to the presence of the threaded rod which transfers a significant amount of heat to the bottom side via conduction.

5.5 Comparison of the TN and FEM Analyses

To validate the results of a simulation, the best option is to have reliable measurement data. Unfortunately, the exact temperature map and thermal conditions of the vacuum chamber were not available, therefore, only rough estimations could be made which are omitted now. However, the principal tests in vacuum were succeed. Consequently, the temperature data on orbit will provide the most accurate information on the quality of the used estimations.

Besides TN modeling, FEM analysis was also performed to check the TN model as their mathematical methods are different, and the number of nodes is significantly higher in the latter case. Figure 7 shows the results in the case of the battery for the two methods. After five orbits, there is 5 °C offset between the two results.

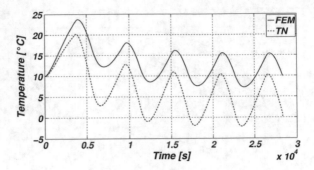

Fig. 7 The battery temperature history at 1367 W/m² solar flux [36]. With permission

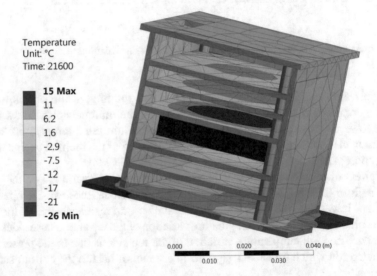

Fig. 8 The global temperature distribution after six hours simulation time (which was a maximum temperature peak) at 1367 W/m² solar flux [36]. With permission

The principal reason behind the difference between the two models is that a single PCB is calculated by FEM in three points along its thickness, and the planar resolution is significantly higher. However, the TN model considered one side as one node. An instantaneous result at the temperature peak of the orbit is shown in Fig. 8. The effect of the battery and the power amplifier on COM is observable on all PCBs. Therefore, if one wishes to improve the accuracy of the TN model, dividing each plane into more components is the best way to do so. It can be done based on a FEM simulation of a similar geometry, following the thermal gradients. The result that a concentrated parameter-based approach is competitive to FEM analysis was also concluded by Corpino et al. [41]. The reason for the lower calculated temperature values is that there is zero inner thermal resistance in a single PCB, leading to faster heat release. Hence, the TN method for evaluating thermal behavior is a safe approximation.

Fig. 9 The top side
temperature history at
1367 W/m² solar flux [36].
With permission

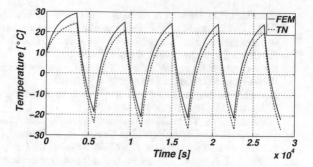

Fig. 10 The bottom side
temperature history at
1367 W/m² solar flux [36].
With permission

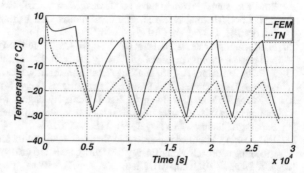

Similarly, 5 °C temperature difference was present in the case of the simulation
of the top panel, shown in Fig. 9. Nevertheless, the temperature deviation between
the two models for the other panels was usually higher due to the above-detailed
reasons. The temperature variation of the bottom panel is characterized by the highest
deviation, which meant 3–16 °C temperature difference, shown in Fig. 10.

This example calculation of a small-sized satellite has shown that an appropriate
TN model can reproduce the characteristics of a FEM model which is rather easy to
establish today. However, design and long-term use calculations are easier with TN
due to the significantly shorter simulation time. Even though it was not emphasized
here, the model is the most sensitive to the boundary and inner conditions due to the
radiative heat transfer between the satellite and the cold space. Therefore, it is advised
to spend enough time on analyzing your system and its properties before running
any model to get dependable results. As a final sentence about the two models, the
TN analysis provides a safe approach compared to FEM due to the heat propagation
inside one node is neglected.

References

1. M. Kociak, L.F. Zagonel, Cathodoluminescence in the scanning transmission electron microscope. Ultramicroscopy **176**, 112–131 (2017)

2. N. Brodusch, H. Demers, A. Gellé, A. Moores, R. Gauvin, Electron energy-loss spectroscopy (EELS) with a cold-field emission scanning electron microscope at low accelerating voltage in transmission mode. Ultramicroscopy **203**(Dec. 2018), 21–36 (2019)

3. P. Mølgaard Mortensen, T.W. Hansen, J. Birkedal Wagner, A. Degn Jensen, Modeling of temperature profiles in an environmental transmission electron microscope using computational fluid dynamics. Ultramicroscopy **152**, 1–9 (2015)

4. S. Gomès, A. Assy, P.O. Chapuis, Scanning thermal microscopy: a review. Phys. Status Solidi A Appl. Mater. Sci. **212**(3), 477–494 (2015)

5. K. Kim, B. Song, V. Fernández-Hurtado, W. Lee, W. Jeong, L. Cui, D. Thompson, J. Feist, M.T. Homer Reid, F.J. García-Vidal, J.C. Cuevas, E. Meyhofer, P. Reddy, Radiative heat transfer in the extreme near field. Nature **528**(7582), 387–391 (2015)

6. Y. Wang, M. Xingsen, S. Shen, W. Zhang, Heat transfer characteristics of steam condensation flow in vacuum horizontal tube. Int. J. Heat Mass Transf. **108**, 128–135 (2017)

7. S.H. Choi, Thermal type seawater desalination with barometric vacuum and solar energy. Energy **141**, 1332–1349 (2017)

8. N. Myneni, A. Date, M. Ward, P. Gokhale, M. Gay, Combined thermoelectric power generation and passive vacuum desalination. Energy Procedia **110**(December 2016), 262–267 (2017)

9. J. Ruan, J. Huang, B. Qin, L. Dong, Heat transfer in vacuum pyrolysis of decomposing hazardous plastic wastes. ACS Sustain. Chem. Eng. **6**(4), 5424–5430 (2018)

10. X. Zhang, D. Huang, W. Jiang, G. Zha, J. Deng, P. Deng, X. Kong, D. Liu, Selective separation and recovery of rare metals by vulcanization-vacuum distillation of cadmium telluride waste. Sep. Purif. Technol. **230**(July 2019), 115864 (2020)

11. L.E. Juanicó, Modified vacuum tubes for overheating limitation of solar collectors: a dynamical modeling approach. Solar Energy **171**(July), 804–810 (2018)

12. T.Y. Wang, Y.H. Zhao, Y.H. Diao, R.Y. Ren, Z.Y. Wang, Performance of a new type of solar air collector with transparent-vacuum glass tube based on micro-heat pipe arrays. Energy **177**, 16–28 (2019)

13. X. Huang, Q. Wang, H. Yang, S. Zhong, D. Jiao, K. Zhang, M. Li, G. Pei, Theoretical and experimental studies of impacts of heat shields on heat pipe evacuated tube solar collector. Renew. Energy **138**, 999–1009 (2019)

14. C. Strauß, R. Gustus, W. Maus-Friedrichs, S. Schöler, U. Holländer, K. Möhwald, Influence of atmosphere during vacuum heat treatment of stainless steels AISI 304 and 446. J. Mater. Process. Technol. **264**(August 2018), 1–9 (2019)

15. M. Asemi, M. Ahmadi, M. Ghanaatshoar, Preparation of highly conducting Al-doped ZnO target by vacuum heat-treatment for thin film solar cell applications. Ceram. Int. **44**(11), 12862–12868 (2018)

16. T.M. Flynn, *Cryogenic Engineering*, 2nd edn. (CRC Press, Taylor & Francis Group, Louisville, CO, 2004)

17. J.P. Holman (ed.), *Heat Transfer*, McGraw-Hill Series in Mechanical Engineering, 10th edn. (McGraw-Hill Education, New York, 2009)

18. M. Martin, K. Holge, *VDI Heat Atlas*, 2nd edn. (Springer, Berlin, 2010)

19. M. Alam, H. Singh, M.C. Limbachiya, Vacuum insulation panels (vips) for building construction industry—a review of the contemporary developments and future directions. Appl. Energy **88**(11), 3592–3602 (2011)

20. A. Sandá, S.L. Moya, L. Valenzuela, Modelling and simulation tools for direct steam generation in parabolic-trough solar collectors: a review. Renew. Sustain. Energy Rev. **113**(June), 109226 (2019)

21. S. Sobhansarbandi, P.M. Martinez, A. Papadimitratos, A. Zakhidov, F. Hassanipour, Evacuated tube solar collector with multifunctional absorber layers. Solar Energy **146**, 342–350 (2017)

22. H.J. Song, W. Zhang, Y.Q. Li, Z.W. Yang, A.B. Ming, Exergy analysis and parameter optimization of heat pipe receiver with integrated latent heat thermal energy storage for space station in charging process. Appl. Therm. Eng. **119**, 304–311 (2017)

23. J. Yang, T. Caillat, Thermoelectric materials for space. MRS Bull. **31**(3), 224–229 (2006)

24. A.D. Kraus, A. Aziz, J. Welty, *Extended Surface Heat Transfer* (Wiley, Hoboken, 2001)

25. F. Hajabdollahi, H.H. Rafsanjani, Z. Hajabdollahi, Y. Hamidi, Multi-objective optimization of pin fin to determine the optimal fin geometry using genetic algorithm. Appl. Math. Model. **36**(1), 244–254 (2012)

26. H. Azarkish, S.M.H. Sarvari, A. Behzadmehr, Optimum design of a longitudinal fin array with convection and radiation heat transfer using a genetic algorithm. Int. J. Therm. Sci. **49**(11), 2222–2229 (2010)

27. R. Das, Estimation of parameters in a fin with temperature-dependent thermal conductivity and radiation. Proc. Inst. Mech. Eng. Part E J. Process. Mech. Eng. **230**(6), 474–485 (2016)

28. C. Wang, J. Chen, S. Qiu, W. Tian, D. Zhang, G.H. Su, Performance analysis of heat pipe radiator unit for space nuclear power reactor. Ann. Nucl. Energy **103**, 74–84 (2017)

29. J. Bouwmeester, J. Guo, Survey of worldwide pico- and nanosatellite missions, distributions and subsystem technology. Acta Astronaut. **67**(7–8), 854–862 (2010)

30. M. Tolmasoff, C. Venturini, Improving mission success of cubeSats, in *Proceedings of the U.S. Space Program Mission Assurance Improvement Workshop* (El Segundo, CA, 2017)

31. C.A. Belk, J.H. Robinson, M.B. Alexander, W.J. Cooke, S.D. Pavelitz, *Meteoroids and Orbital Debris: Effects on Spacecraft*, Technical Report, NASA Marshall Space Flight Cente, Huntsville, AL (1997)

32. S.B. Khan, A. Francesconi, C. Giacomuzzo, E.C. Lorenzini, Survivability to orbital debris of tape tethers for end-of-life spacecraft de-orbiting. Aerosp. Sci. Technol. **52**, 167–172 (2016)

33. T. Maury, P. Loubet, M. Trisolini, A. Gallice, G. Sonnemann, C. Colombo, Assessing the impact of space debris on orbital resource in life cycle assessment: a proposed method and case study. Sci. Total Environ. **667**, 780–791 (2019)

34. S. Lee, A. Hutputanasin, A. Toorian, W. Lan, R. Munakata, J. Carnahan, D. Pignatelli, A. Mehrparvar. *CubeSat Design Specification Rev. 13*, (2014)

35. D.G. Gilmore, *Spacecraft Thermal Control Handbook*, 2nd edn. (Aerospace Press, El Segundo, CA, 2002)

36. R. Kovács, V. Józsa, Thermal analysis of the SMOG-1 PocketQube satellite. Appl. Therm. Eng. **139**, 506–513 (2018)

37. W. Stenzel, *A Sunny Outlook for NASA Kepler's Second Light*, Technical report, NASA Ames Research Center, Moffett Field, CA (2013)

38. N.D. Anh, N.N. Hieu, P.N. Chung, N.T. Ahn, Thermal radiation analysis for small satellites with single-node model using techniques of equivalent linearization. Appl. Therm. Eng. **94**, 607–614 (2016)

39. M. Bonnici, P. Mollicone, M. Fenech, M.A. Azzopardi, Analytical and numerical models for thermal related design of a new pico-satellite. Appl. Therm. Eng. **159**, 113908 (2019)

40. A. Torres, D. Mishkinis, T. Kaya, Mathematical modeling of a new satellite thermal architecture system connecting the east and west radiator panels and flight performance prediction. Appl. Therm. Eng. **65**(1–2), 623–632 (2014)

41. S. Corpino, M. Caldera, F. Nichele, M. Masoero, N. Viola, Thermal design and analysis of a nanosatellite in low earth orbit. Acta Astronaut. **115**, 247–261 (2015)

Chapter 5
Nature Knows Better

1 Introduction

The classical constitutive equation of heat conduction is valid for homogeneous materials; moreover, the 'rapidity' of a process does matter as well, i.e., a microsecond fast process may not be modeled appropriately with the Fourier's law. There are analogous restrictions for the other constitutive equations; however, these limits cannot be determined easily. The thermodynamical background of constitutive laws is discussed in Chap. 2. Then we show here the experimental background which acted as a motivation to find a better model, and also, to decide which model is applicable and what are their limits. All these phenomena are beyond the classical modeling capabilities from some sense. The validity limit of each constitutive law is possible to characterize with certain time and length scales. These scales arise from various sources, for instance,

- the material parameters which determine the observed propagation speeds,
- the structure of the material,
- the size of the conducting medium,
- the boundary conditions,

and, implicitly, the actual state of the material. Now, let us recall the relevant material parameters for heat conduction equations:

- the classical coefficients are the mass density ρ, the specific heat c and the thermal conductivity λ. These form the so-called thermal diffusivity $\alpha = \lambda/(\rho c)$ which combination is essential for transient (time-dependent) problems. Usually, it falls between 10^{-8} and 10^{-5} m^2/s, depending on the conductivity ability and the specific heat capacity (ρc) of the medium. This is generally true if the temperature is around $-40-200\,^\circ$C as it is in most of the engineering applications.
  However, one must be more careful with low or high-temperature situations such as the modeling of burning processes and its counterpart, the cryogenic applications.

© Springer Nature Switzerland AG 2020
V. Józsa and R. Kovács, *Solving Problems in Thermal Engineering*, Power Systems,
https://doi.org/10.1007/978-3-030-33475-8_5

The material parameters depend on the state variables,[1] most commonly on the temperature [1, 2], its order of magnitude is also a characteristic property of the material.

- If one have to apply a generalization of the constitutive equation, then the new coefficients introduce further time and length scales. One outstanding example is related to the Guyer-Krumhansl equation. After eliminating the heat flux from the equations,[2] it reads

$$\tau \partial_{tt} T + \partial_t T = \alpha \Delta T + l^2 \Delta \partial_t T, \tag{1}$$

in which the ratio of l^2/τ has the same dimension as α. Moreover, an attentive reader notes the appearance of the Fourier equation and also its time derivative with the new parameters. Thus the relation between α and l^2/τ can characterize the deviation from the Fourier heat equation. In the special case of $\alpha = l^2/\tau$, the solution of the Fourier equation is recovered and called 'Fourier resonance condition' [3, 4]. Furthermore, the appearance of l^2/τ introduces a new time scale beside the classical one. It is worth to mention that l^2 itself has the dimension of m^2 which proposes some length scale dependence, too.

That structure of the Guyer-Krumhansl equation suggests parallel conduction phenomena with separate time and length scales. This usually happens in heterogeneous materials. The experimental section about the room temperature measurements presents the related observations.

- The new parameters in the constitutive equations can depend on the state variables. The temperature dependence of the relaxation time is experimentally proved in low-temperature situations.
- The length scaling properties become relevant in micro and nano-sized objects.[3] According to the related experiments, some of the material coefficients reflect their dependence on body size. Moreover, it is also observable on macroscale in certain heterogeneous materials.
- The role of the boundary conditions is not completely clear from the sense of observing non-classical phenomena. In most cases, the so-called heat pulse experiment [5–7] (or flash experiment) is applied since the time scale of the heat pulse can be varied in a wide range, e.g., from 1 μs to seconds. Its schematic arrangement is depicted in Fig. 1. The observation of non-Fourier phenomena depends on both the excitation frequency (even if it is periodical or a single pulse) and the heat loss to the environment.

First, we start with the heat conduction experiments, beginning at the low-temperature cases as their historical and theoretical importance is still outstanding.

---

[1] Since the heat flux $\mathbf{q}$ becomes a state variable in the generalized equations, independently of the approach, it is worth to mention the possible $\mathbf{q}$-dependence of the material parameters.

[2] We refer to it as 'temperature representation'. If the temperature is eliminated, then it is called 'heat flux representation'.

[3] Naturally, the scaling property appears in large scales, too, but it is less frequent in the practical applications and may be important merely for geological applications.

Fig. 1 Schematic arrangement of a heat pulse experiment [8]. On the left side, a heat pulse irradiates the front face, and, in the meanwhile, the temperature is measured at the rear side

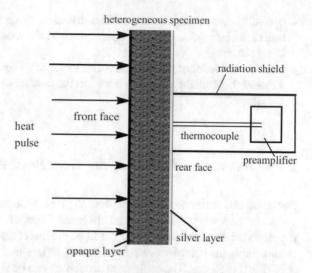

heterogeneous specimen

radiation shield

heat pulse

front face

thermocouple

rear face

preamplifier

silver layer

opaque layer

That section starts with the observations of second sound, the damped wave propagation of heat and ends with the ballistic type conduction, which represents the connection between the macroscale low-temperature and the nanoscale room temperature problems. Later on, further room temperature experiments are discussed, including examples for non-Fickian and non-Newtonian phenomena.

Now, for the sake of clarity, let us summarize the relevant heat conduction models from Chap. 2. All these models have importance in the experiments.

- Fourier's law: $\mathbf{q} = -\lambda \nabla T$.
- Maxwell-Cattaneo-Vernotte (MCV) equation: $\tau \partial_t \mathbf{q} + \mathbf{q} = -\lambda \nabla T$.
- Guyer-Krumhansl (GK) equation: $\tau \partial_t \mathbf{q} + \mathbf{q} = -\lambda \nabla T + l^2 \Delta \mathbf{q}$.
- Dual Phase Lag (DPL) equation: $\mathbf{q}(\mathbf{x}, t + \tau_q) = -\lambda \nabla T(\mathbf{x}, t + \tau_T)$.
- Ballistic-conductive (BC) equations: $\tau_q \partial_t \mathbf{q} + \mathbf{q} + \lambda \nabla T + l \nabla \cdot \mathbf{Q} = 0$ and $\tau_Q \partial_t \mathbf{Q} + \mathbf{Q} + l \nabla q = 0$.

2 Non-Fourier Heat Conduction

Previously, we emphasized some general aspects of the validity region of constitutive equations. In the following, we shall focus on the phenomena that cannot be explained with Fourier's law. Their experimental background is extensive; thus, it is not possible to present them in detail. However, this summary is enough to highlight the scaling properties of the heat conduction phenomenon. In order to gain a better overview of the observed phenomena, we first briefly summarize the existing propagation modes:

- diffusive propagation: the Fourier's law is applicable;
- second sound: damped wave propagation, slower than the speed of sound, the MCV or GK equations are necessary to use;

- over-diffusive propagation: it is also diffusive conduction with different characteristics, observed on room temperature, and the modeling level of GK equation is necessary;
- ballistic conduction: propagates with the speed of sound, its description requires a model that contains coupling between the heat flux and the pressure, for details, see Chap. 2.

2.1 Classical Evaluation of the Heat Pulse Experiments

For comparative reasons, it is expedient to present the classical case, too. In Fig. 2, a common temperature history can be observed. There are two possibilities to evaluate the recorded data. The first choice could be the direct simulation of the experimental setting using the complete system of partial differential equations. This is also presented in this Figure. The evaluation procedure can be significantly simplified if one presumably accepts the validity of the Fourier's law: the given analytical solution[4] immediately offers the thermal diffusivity, the outcome of this measurement [7]:

$$\alpha = 1.38 \frac{\delta^2}{\pi^2 t_{1/2}}, \tag{2}$$

where $t_{1/2}$ is the time required to reach half of the asymptotic temperature value that is given from the data and δ stands for the sample thickness. Moreover, this solution also assumes that the first term in the infinite sum (for details, see [9–11]) is sufficient.

In simple cases, it is easy to use and yields accurate results. However, for non-Fourier problems, such solutions are still under development, and their practical usage is not straightforward [12–16]. Furthermore, the deep understanding of the role of

Fig. 2 Typical rear side (dimensionless) temperature history [17]. Solid line: filtered experimental data. Dashed line: prediction of Fourier equation. The value 1 represents the adiabatic limit. Since cooling is present, the dimensionless temperature remains below 1

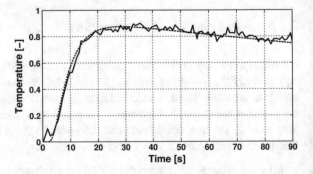

---

[4]Assuming adiabatic boundary conditions on both sides with non-uniform initial condition to model the initial (and quick) absorption of the heat pulse.

constitutive equations is required to define the boundary conditions appropriately. Their role will be discussed later. Since for these more complicated cases, there is no applicable formula, one way remains: the direct numerical solution of the corresponding PDE system.

2.2 Low-Temperature Heat Conduction: Second Sound

The history of non-Fourier heat conduction has been started with predicting the phenomenon of second sound, a dissipative, wave-like propagation of heat, by Tisza and Landau [18–21]. During that era, the second sound was assumed to be present in a superfluid medium such as the He behaves around 1.5 K. Describing the behavior of a superfluid is still the subject of recent researches. In modeling a superfluid, the pioneering idea of Tisza was to assume the mixture of a superfluid (Bose-Einstein condensate) and 'non-condensate' parts, Tisza developed the first basis of a two-fluid model [22]. The existence of second sound is first experimentally observed by Peshkov in liquid (superfluid) He II in 1944 [23]. In parallel, the behavior of liquid He—its phase diagram, viscosity properties, vorticity formations and their impact on thermal processes—have been investigated widely and are still important questions [24–35].

The wave propagation speed plays a crucial role in many aspects. On the one hand, it serves as a guide what has to be measured, what is the time scale. On the other hand, it helps to identify the signal by distinguishing it from the other wave phenomena. One example is shown in Fig. 3, where the time scales are also visible. Here, the waveform of heat conduction became apparent, and in some cases, its reflection is also observed [36].

Here, we mention the work of Lane et al. [38, 39], Maurer and Herlin [40], Ward and Wilks [41], Atkinson and Osborn [42], Chester [43] and Pellam [36], who analyzed theoretically and experimentally that attribute of the second sound. The measurements of Pellam begin at the λ-point of He II (around 2.2 K), observing about 2 m/s propagation speed. Moreover, these observations clearly present the nonlinear behavior of temperature dependence: decreasing the temperature by 0.3 K, the speed increases slightly above 20 m/s [36]. This is in agreement with the work of Lane et al., who reproduced the experimental results of Peshkov with greater accuracy. According to their measurements, the speed of second sound is 19.18 ± 0.1 m/s at 1.74 K. Atkinson and Osborn extended these results, they performed measurements on He II at 0.1 K, and the speed found to be 152 ± 5 m/s, see Fig. 4 for details. It shows that the speed starts to increase considerably as approaching 0 K.

Later, Guyer and Krumhansl obtained the so-called window condition [44], an interval to estimate which heat pulse frequency can create a measurable heat wave. It is proved to be extremely helpful to detect second sound in solids as well [37, 45–48] and plays an important role in recent studies [12–14, 49, 50]. An overview of the history of this topic is given by Joseph and Preziosi, Ván, Cimmelli and coworkers [51–59].

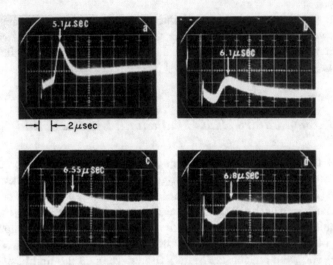

FIG. 9. Typical heat pulses in Bi III. Propagation
length 5.1 mm. The pulses broaden and move to later
time as temperature T is raised. Curve (a) 1.95 K; (b)
2.68 K; (c) 3.1 K; (d) 3.3 K. Propagation along C_3 axis.

Fig. 3 Typical temperature history in detection of heat pulses [37]. With permission

Fig. 4 The temperature
dependence of the
propagation speed of second
sound in He II based on [42].
With permission

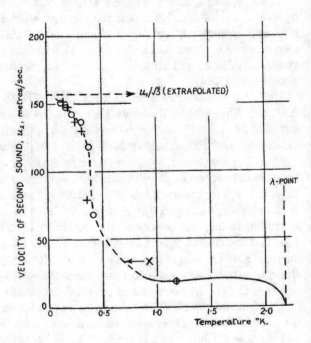

The first extension of Fourier's law, the so-called MCV equation is devoted to model the phenomenon of the second sound. It characterizes the propagation speed with the formula

$$v_c(T) = \sqrt{\frac{\alpha}{\tau}}(T),$$ (3)

in which the temperature dependence of the relaxation time can be recovered if the function $\alpha(T)$ is also measured. Since $v_c(T)$ is nonlinear, together with $\alpha(T)$, thus $\tau(T)$ could be nonlinear, too, especially close to 0 K. We recall that the specific heat and the mass density are also the functions of the temperature. Eventually, the measured $v_c(T)$ reflects T-dependence of 4 (not necessarily) independent material parameters. Hence various T-dependency can be imagined for $\tau(T)$. One of the simplest example is

$$\tau(T) = 3.5\text{ps} - \frac{2\text{ps}}{500\text{K}}(T - 300\text{K}),$$ (4)

which holds in a certain temperature range (< 1000 K) for graphene with T being the lattice temperature [60].

2.3 Low-Temperature Heat Conduction: Ballistic Propagation

The next breakthrough in experimental heat conduction was discovering the ballistic-type propagation. Jackson, Walker and McNelly performed several experiments in which they observed a signal with the speed of sound [61–64]. In Fig. 5, typical temperature histories are depicted, including also the second sound. The parallel appearance of different phenomena makes this experiment outstanding. Here, data recorded for two NaF samples are presented, where L and T denote the longitudinal and transversal ballistic signal, respectively. Those are faster than the second sound.

One essential requirement for successful measurement is having an isotopically and chemically pure crystal, which is challenging to produce [64]. These anisotropic crystals have incredibly high thermal conductivity, between 1000 and 50000 W/(mK) which is extremely sensitive to the temperature and also depends on the direction [65–71].

Various theories have been developed to provide an explanation, the most important ones are related to Rogers [72, 73], Dreyer and Struchtrup [74], Frischmuth and Cimmelli [75–77], Chen [78, 79], Ma [80–82], Jou et al. [83–89], Kovács and Ván [90–92].

Rogers' model originates from Landau's [93] in which Landau assumed that the bulk (or second) viscosity η_b of a fluid is complex in case of sound propagation, i.e.,

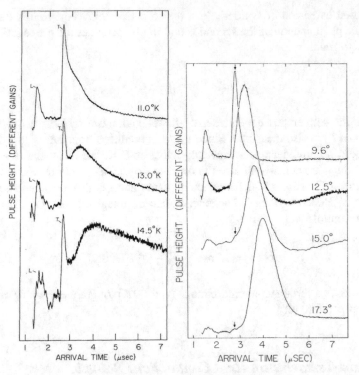

Fig. 5 Original results of NaF experiments: the right one was published in [61] and the left one in [63]. L and T denote the longitudinal and transversal peaks of ballistic propagation, respectively. The length of a heat pulse is around 0.1 μs. The sample length is around 7 mm. With permission

$$\eta_b = \frac{e\tau(1 - c_2^2/c_1^2)}{1 - i\omega\tau}, \tag{5}$$

where $\tau^{-1} = \tau_R^{-1} + \tau_N^{-1}$ with the corresponding relaxation times of resistive and normal processes which are introduced earlier in Chap. 2. Furthermore, ω, c_1 and c_2 are the frequency, the first and the second sound velocities, respectively; e is the internal energy that is splitted into an equilibrium e_0 and a perturbated e' part:

$$e = e_0 + e'. \tag{6}$$

Rogers adopts that approach, and the hydrodynamic equations are applied for phonon gas in which the heat flux vector replaces the velocity, $\mathbf{q} = e\mathbf{v}$. Rogers has shown the transition from second to the first sound wherein the frequency of normal processes together with the bulk viscosity have great influence. It is also important to note, that model fits well for second sound measurement data. However, it is not able to produce temperature history showing the ballistic and second sound propagation together. Although Y. Ma improved the complex viscosity model of Rogers, it is still not able to give an appropriate explanation for NaF experiments of McNelly et al.

In the works of Frischmuth and Cimmelli, the phonon model was discarded and a so-called "semi-empirical" approach have been applied. Here, a non-equilibrium temperature β is introduced beside the absolute one—denoted by Θ—and a single relaxation equation is assumed to be valid, i.e.,

$$\tau d_t \beta + \beta = \Theta. \tag{7}$$

Here, β behaves as a single scalar internal variable that accounted as follows,

$$\mathbf{q} = -\lambda(\beta, \Theta)\nabla\beta, \tag{8}$$

that is, the gradient of non-equilibrium temperature appears in this modified Fourier's law. That model is proved to be useful in describing the second sound, presenting an alternative way to include the wave nature of heat conduction into the modeling. Later, this semi-empirical description is extended by including the mechanical contributions, too.

The previous approach is analogous with the concept of the micro-temperature model developed by Berezovski et al. [94–97]. They assumed that exists a non-equilibrium variable, the so-called micro-temperature, which is coupled to the macro-deformation. Then, the change in the micro-temperature induces a perturbation on macro-temperature (that is modeled with the Fourier's law), too, which results in a wave propagation of heat with finite speed, limited by the mechanical properties of the medium. Defining a convenient free energy function, all these effects can be included [94]. The approach of dual internal variables motivates the form of such free energy, see the work of Ván, Berezovski et al. [96, 98, 99]. Despite the missing experimental benchmarks, this approach concludes that there is no need to modify the Fourier's law in order to obtain a heat wave propagating with finite speed.[5]

The model of Dreyer and Struchtrup [74, 100] originates in phonon hydrodynamics, but it differs from Rogers' approach. This is the first model that contains the ballistic and second sound effects in parallel, with a quantitative match to the experiments. The thermo-mechanical coupling arises naturally from the momentum series expansion of the Boltzmann equation, and it also laid the foundation for modeling such a complex problem. Its drawbacks have mostly mathematical origin: the propagation speed is not accurate using only 3 momentum equations. Besides, there is no solution to this shortcoming within this framework since the relevant coefficients are fixed by definition.

The NaF experiments are first quantitatively reproduced using the approach of non-equilibrium thermodynamics with internal variables [90, 97, 101–103], which is presented in Chap. 2. The corresponding solutions of the ballistic-conductive model are presented in Fig. 6 [92].

---

[5] As a philosophical question to the Reader: is it possible to interpret the internal variable as a measure of deformation to obtain a hyperbolic heat equation?

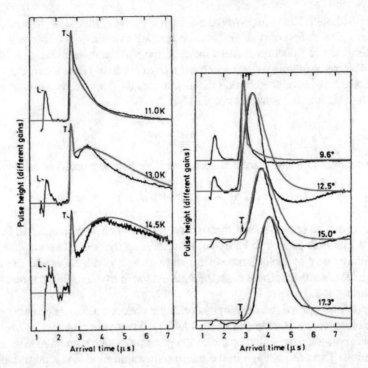

Fig. 6 Summarizing the calculations and the experiments, the red curves correspond to the solutions of the one-dimensional form of ballistic-conductive model [92]. With permission

Fig. 7 Temperature dependence of relaxation times for a NaF crystal, evaluated using the ballistic conductive model [92]. It corresponds to the sample #607167J [64], (left side in Fig. 5). With permission

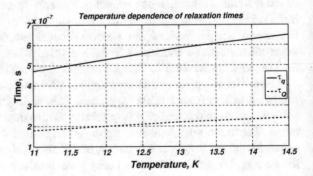

For detailed comparison with phonon hydrodynamics, see [91, 92, 104]. These simulations predict the temperature dependency of relaxation times (see Fig. 7), too. In each case, the relaxation times are fitted and the material parameters (thermal conductivity, specific heat and mass density) are taken from earlier measurements [46, 63, 105–107], summarized in the Tables 1 and 2. The mass density was constant since there is no information about its T-dependence for NaF crystals.

Table 1 Classical material parameters for crystal #607167J, (left side in Fig. 5)

	Thermal conductivity $\left(\frac{W}{mK}\right)$	Specific heat $\left(\frac{J}{kgK}\right)$	Mass density $\left(\frac{kg}{m^3}\right)$
@11 K	8573	1.118	2866
@13 K	10200	1.8	2866
@14.5 K	10950	2.543	2866

Table 2 Classical material parameters for crystal #7204205W, (right side in Fig. 5)

	Thermal conductivity $\left(\frac{W}{mK}\right)$	Specific heat $\left(\frac{J}{kgK}\right)$	Mass density $\left(\frac{kg}{m^3}\right)$
@9.6 K	8500	0.7123	2866
@12.5 K	17300	1.62	2866
@15 K	21750	2.735	2866
@17.3 K	22880	4.45	2866

2.4 Ballistic Conduction: Observations on Room Temperature

It is useful if we insist on the kinetic definition of ballistic propagation: phonons scatter on the boundary only. For a low-temperature system, the mean free path of a particle becomes large enough to make the ballistic contribution significant in the overall heat conduction process. This is analogous if the system size is reduced instead of the temperature, this effect becomes apparent in a nanoscale object.

Besides, the determination of material parameters also becomes difficult from an experimental point of view. For example, the size dependence of thermal conductivity is also observed in several cases, e.g., for superlattices, beyond the usual temperature dependence [83, 108–118]. Wang et al. [119] reported results about the strong size and temperature dependence of thermal conductivity, observed in nanofilms and nanotubes by several authors [110, 120–125], and also for semiconductors [126, 127]. In the case of modeling nanosized materials, the molecular dynamics simulations have greater importance [110, 119, 128].

Usually, the non-Fourier heat conduction can be detected in a different way, such as measuring the conductivity properties. Instead of various wave propagation modes, the thermal and electric resistance measurements show the deviation from the classical theory [129], see Fig. 8 in which an example is shown about the thermal behavior of carbon nanotubes.

However, there is one outstanding example for nanomaterials, in which case the heat pulse experiments can be conducted similarly as previously. The research about the thermal behavior of thin films is still increasing [130]. The numerous applications and processes [131, 132] all require the reliable prediction of temperature conditions.

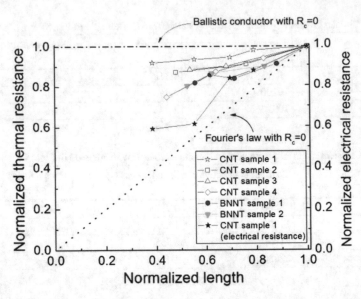

Fig. 8 Thermal resistance measurements show the presence of ballistic propagation [129]. Abbreviations: CNT (carbon nanotubes), BNNT (boron-nitride nanotubes). With permission

From a theoretical point of view, a continuum model remains applicable, but it is still not clear which one would be the best one in such cases. Using a kinetic description is more popular, and regarding the particular heat transport mechanism, shows good agreement both with molecular dynamics simulations and experiments [133, 134]. For more in-depth insight, we recommend the work of Chen [78, 79, 135–137].

In this experimental situation, the pulse length becomes significantly shorter than before; it is in order of femtoseconds, e.g., $100-200$ fs. The usual thickness is around $200-2000$ Å [138]. Figure 9 shows the usual arrangement of a heat pulse experiment on a nanofilm material. It is important to emphasize that instead of measuring the temperature, the change of reflectivity is detected; for instance, see the data in Fig. 10 [139]. It is possible to reconstruct the temperature from that reflectivity change; moreover, the characteristics of the transient behavior of that change depends on the actual mean free path in the film [140]. That change of reflectivity is proportional to the temperature, and that depends on the material. Recently, the group of Siemens, Hoogeboom-Pot and Lee et al. [141–143] obtained convincing experimental proof of the ballistic behavior at room temperature.

2.5 Room Temperature Heat Conduction: Over-Diffusion

In the previous sections, we have seen various forms of non-Fourier heat conduction, namely, the second sound and ballistic propagation. From the kinetic point of view,

Fig. 9 The usual experimental arrangement for thin nanofilms [138]. With permission

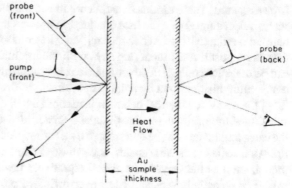

probe
(front)
pump
(front)
probe
(back)
Heat
Flow
Au
sample
thickness
FIG. 1. Schematic of the experimental setup. The sample is pumped on the front surface, and the change in reflectivity at either the front or the back surface is probed.

Fig. 10 The recorded change in resistivity in respect to time for various materials [139]. With permission

Pb
W
Cr
V
Nb
NbN
ΔR (arb. scale)
-0.4
0
0.4
0.8
1.2
Delay (ps)

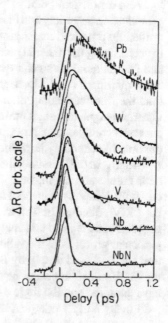

these are the consequence of large mean free path that also can be observed in a nanoscale object. The Knudsen number $\mathrm{Kn} = l/L$ reflects this effect very well. While in the low-temperature situation the mean free path l becomes large, comparable with the body size L, in the room-temperature case the body size L is decreased to the order of l.

This section, however, differs from the previous ones. It is not widely known that there is one more non-Fourier phenomenon, called over-diffusive propagation. The classical law of heat conduction is valid for homogeneous materials which practically do not exist in reality. Hence in this section, the behavior of heterogeneous materials is investigated. Their various constituents introduce different time scales (or spatial scales, accordingly) into the heat conduction process due to the presence of different thermal conductivities and heat capacities. The consequence of the thermal interaction among each constituent (together with the thermal resistance on the interfaces) can be measured only 'effectively' since it is practically not possible to measure the temperature history point by point. That 'effective' detection means a 'homogenization', the measured thermal behavior is reconstructed using a generalized model that substitutes the original—heterogeneous—material with a homogeneous one. Then the extra parameters can be determined by a fitting procedure, assuming that the multiple timescales (or spatial scales, accordingly) can be described with a generalized model, such as the Guyer-Krumhansl equation.[6] One particular set of parameters, $\alpha < l^2/\tau$, is called over-damped or over-diffusive propagation. This is observed in several rock and foam samples, which is presented later. The opposite relation has not been measured so far.

From a practical point of view, there is a difficulty in the treatment of heterogeneities. Their variety and distribution are different from sample to sample even for artificially produced materials such as foams. Consequently, there is a random error in determining the material parameters [144, 145], even for the same type of material and that material parameter can only be the 'effective' (or apparent) one since this is the measured one.[7] Moreover, these materials are anisotropic, i.e., their behavior (and their corresponding time scales, accordingly) is direction-dependent, and the material parameters are represented by tensors instead of scalars. Furthermore, they possess a complex structure (for instance, a porous one) which is also practically not possible to accurately implement in the modeling process.

Usually, in the literature, the MCV equation is considered to be valid also on room temperature and attempted to be found in the aforementioned experiments unsuccessfully. Despite the inconclusive experimental results, Bamdad et al. [146] applied the MCV equation for heat transfer via fins which fin was assumed to be homogeneous. In this regard, this analysis has remained on the level of theory without any particular proof. Both the validity and the misinterpretations around the MCV equation are discussed deeply by Bright and Zhang [147]. According to their argument, the MCV equation cannot be observed in any room temperature experiment. They have the same opinion for low-temperature experiments, too, which is contradictory with many early works that used the MCV model[8] successfully in describing the phenomenon of the second sound.

---

[6]Recall the Fourier resonance condition!

[7]We note here the difference between the apparent and theoretical (calculated) material parameters. It has great importance in some scaling properties that will be discussed at the end of this chapter.

[8]In some papers, the MCV equation is considered to be thermodynamically incompatible. In general, this interpretation depends on the space of state variables: one must include the heat flux as an independent variable, then the anomalies around the entropy production disappear.

Porous Materials

The porous material structure is advantageous for many practical applications, and not merely in civil engineering. The possibilities are widened and extended to the field of mechanical engineering, too. Naturally, the complete description of a hetero-geneous material requires many physical parameters [145]. The most relevant one is the so-called 'porosity', $\mathcal{P}$ that defines the volume ratio between the voids and the apparent volume. Its presence modifies both the mechanical and thermal attributes, consequently, the corresponding material parameters depend on $\mathcal{P}$ [148–152]. A detailed description of the transport properties and the related models, especially for fluid flow, are given in [145, 153]. Liu [154] profoundly presents both the math-ematics and physics of porous structures. It is worth to note that the approach of non-equilibrium thermodynamics is mostly out of sight in the corresponding litera-ture and consequently, the Fourier's law is not modified consistently [155].

The history of room temperature non-Fourier heat conduction is not as long as the low-temperature research; experiments in this direction have been conducted for around 50 years. Luikov may be the first who proposed the Maxwell-Cattaneo-Vernotte equation to model a room temperature process in porous materials, in 1965 [156, 157]. The first example to detect room temperature non-Fourier effects is related to Kaminski [158], see Fig. 11 for the setup and the results. The examined materials were sand and glass ballotini, among others. These are grained materials, and the different time scales may exist due to the voids among the particles. It is found that there is a significant time lag between the heat pulse and the signal observation. As Kaminski states, if the Fourier's law would be valid, the signal should arrive between 5 and 10 s. However, the signal was observed much later, between 100 and 200 s (Fig. 11). Unfortunately, these results could not be reproduced later [147, 159–161].

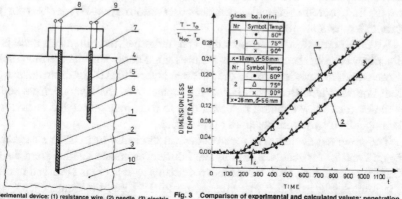

Fig. 1 Experimental device: (1) resistance wire, (2) needle, (3) electric insulation, (4) needle, (5) thermocouple, (6) insulation (7) support, (8) to power supply and stabilizer, (9) to amplifier and recorder, (10) material

Fig. 3 Comparison of experimental and calculated values: penetration time; (1, 2) experimental and calculated values of temperature; (3, 4) penetration time for $x = 18$ mm and $x = 28$ mm layers, respectively

Fig. 11 Left side: the setup of Kaminski's experiment. Right side: the measured temperature signal respect to time, taken from [158]. With permission

Later, Vedavarz et al. [162] investigated the validity range of MCV equation, then together with Mitra [163], they performed the next, contested experiment. Their results show the wave-nature of heat conduction in frozen processed meat. The meat is situated suddenly onto a preheated plate, and the temperature is measured by thermocouples inside. However, this experiment could not be reproduced, too. Other, follow up measurements led to contradictory results, see Scott et al. [164], Herwig and Beckert [160, 161], Bright and Zhang [147], Roetzel et al. [165]. According to Antaki, additional processes and heat transfer mechanisms might have played a role [166], especially phase transitions from ice to liquid.

It is interesting to mention that on the contrary of Kaminski's experiments, Jiang et al. [167] managed to detect the wave form of heat conduction in a porous material at room temperature. Despite the promising result, there is no either reproduction or further development. Their evaluation is also not sufficient; both the MCV and the dual phase lag model failed to match the measurement quantitatively. Other papers report non-Fourier effects on room temperature, e.g., in liquids and gases [168, 169]. Besides, those are based on the widely criticized dual phase lag model that makes their results more debatable in modeling convective heat transfer.

The corresponding literature of room temperature non-Fourier heat conduction in heterogeneous materials with macroscale size is sparse. From a practical point of view, the modeling problems of rocks are also interesting and have a wide range of applicability. For instance, the extensometer measurements conducted in the nuclear waste disposal of Bátaapáti in Hungary [170] show significant temperature effects in rocks. It motivated the research of the group in BME (Hungary) to perform heat pulse experiments on various rock samples such as limestone and leucocratic rocks [3, 4, 8]. The studied specimens are in mm scale. The preliminary measurements are started with a 'book', that is, the experiments are conducted on a periodic layered structure where the layers are perpendicular on the direction of heat flux. The results were not convincing [171]. In the next step, a capacitor sample is used where the periodic layered structure is present again; however, in parallel with the heat flux. It was the first, reproducible and significant observation of non-Fourier heat conduction at room temperature [3].

We note here again, it corresponds to an 'effective' measurement: there is a very thin silver layer at the rear side that ensures good contact to the thermocouples and also averages the temperatures of the different layers. That temperature history differs from Fourier's law. Besides, it is still anticipated that the layers behave separately as the Fourier's law dictates; however, due to their interaction, the behavior of the 'homogenized' structure becomes non-Fourier.

The observed deviation in rocks occurs in the same form as in a capacitor, see Figs. 12 and 13 for details. Recalling the Fourier resonance relation from the Guyer-Krumhansl equation, the over-diffusive domain ($\alpha < l^2/\tau$) is observed in all cases. It is remarkable to notice that after some initial time, the Fourier heat equation becomes applicable, the temperature history is according to Fourier's law. All the experimental data are fitted well with the Guyer-Krumhansl equation. The resonance condition [172] suitably characterizes that deviation. Interestingly, we observed the size dependence of the over-diffusive phenomenon recently [8]. That dependence

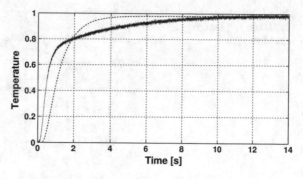

Fig. 12 Measurement results (solid noisy line) and Fourier fitting (dashed line) of crystalline limestone sample from Villány (22/02/2016, second experiment) [104]

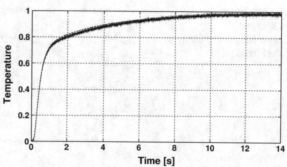

Fig. 13 Measurement results (solid noisy line) and GK fitting (dashed line) of crystalline limestone sample from Villány (22/02/2016, second experiment) [104]

is motivated from a theoretical point of view as well: the change of spatial scale modifies the coefficient l in the GK model. Finally, we note that the derivation of the GK equation in the framework of NET-IV is independent of the material type, i.e., it can be practically applied with no restriction to the structure. Due to the large variety of heterogeneous materials, it can be a significant advantage.

That universality allows applying the same thermodynamic background for metal foams, too [4]. Foams belong to the most pioneering material structures [173–176]. Inclusions are artificially situated inside the material during a special regulated gasifying process. The aim is to preserve good mechanical properties along with a lightened structure. The role of metal foams in aerospace, automotive industry, moreover in the field of heat exchangers is growing [175]. In our heat pulse experiments, two different aluminum-based foams are used, and the same non-Fourier phenomenon is detected, see Figs. 14 and 15 for details.

One excellent example is the modeling task of porous burners in which several transport processes occur at the same time. The voids and the capillaries of the burner are filled with fuel-air mixture which burns in the cavities of the solid material. Its detailed theoretical analysis points beyond the capabilities of the existing theories due to the numerous complexities occurring here. One would have to unify the thermal, the mechanical, the fluid flow, the diffusion, and the chemical equations and aspects which, at that moment, is practically unfeasible. As we have seen previously, these aspects are challenging even separately. This is the reason why an 'effective'

Fig. 14 Measurement results and Fourier fitting of metal foam [104]

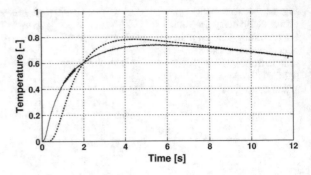

Fig. 15 Measurement results and Guyer-Krumhansl fitting of metal foam [104]

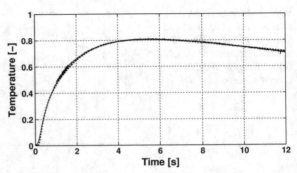

modeling approach could be more useful for some modeling purposes after understanding how the new parameters describe the medium under certain conditions. It still requires much more effort and possesses great potential.

Mujeebu et al. [177, 178] summarizes the simplifying assumptions that usually made in the modeling process:

- the air thermo-physical material properties depend on the temperature and the concentration,
- the solid porous phase described by constant properties,
- the pressure drop inside the burner can be neglected,
- the air and the fuel are completely premixed,
- the flow is considered to be incompressible.

Furthermore, the thermal radiation becomes extremely important that is accounted for the solid phase only; it is considered as a gray body while the gaseous phase is completely transparent. In general, the momentum equation is also out of sight. Consequently, the experimental background is more significant, and there are huge efforts in the literature to develop a reliable model and implement these burners into industrial applications [179–184]. Figure 16 shows a particular porous burner, presenting some characteristics, too.

Completely neglecting the chemical reactions from the mathematical and physical description of porous burners, we arrive at the diffusion (and heat transfer) problem

Fig. 16 The schematic
arrangement of a two-layer
porous burner [177]. With
permission

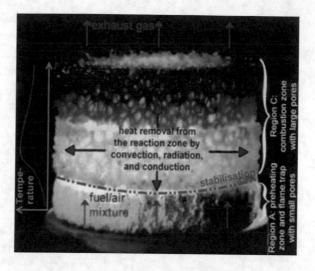

of porous materials. From a practical point of view, this is the modeling task of a
drying process [185–188]; it is important in civil engineering [189] and also occurs
in biological applications. The driving force of the mass flux can be the pressure
gradient,

$$\mathbf{J}_m = -D_m \rho \nabla p, \tag{9}$$

where $\mathbf{J}_m$ being the mass flux, $D_m = k/\mu$ comes from the ratio of the permeability
k and the viscosity μ; this is the so-called Darcy's law [190, 191]. In terms of the
concentration gradient, this is called as Fick's law, which is already presented in
Chap. 2. Analogously to heat conduction, it is not straightforward how to use Fick's
law for porous materials. Unfortunately, the experimental possibilities are limited
since it is more difficult to measure the transient change of the moisture content.

Mathematically, the generalization of Fick's law is analogous to the Fourier's law
[192–194]. Despite the experimental difficulties, Wong et al. [195] reported a series
of experiments in which a disk shape membrane specimen is exposed to a moisture
diffusion. This measurement is designed to be one dimensional. In Fig. 17, we can
observe the appearance of non-Fickian diffusion in a surprisingly similar way as in
case of room temperature non-Fourier heat conduction previously.

The same phenomenon is observed for other compounds, too. According to Wong,
the non-Fickian diffusion is probably due to the parallel diffusion processes, e.g.,
water and vapor. Moreover, the diffusivity of vapor is much higher than that of water.
Adding the fact that all these go on in a porous material, the analogy is almost
complete with non-Fourier heat conduction regarding the parallelly occurring time
and length scales. It is important to emphasize that it depends on the moisture content;
that is, for lower moisture content, the deviation is disappeared.

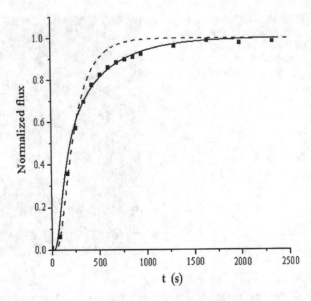

Fig. 17 Measurement results versus a prediction of Fick's law (dashed line) and non-Fickian model (solid line) [196]. With permission

The generalization of Darcy's law is not completely analogous. Originally, the Darcy equation reads

$$\mathbf{v} = -D_m \nabla p, \tag{10}$$

in which the velocity is proportional to the pressure gradient. However, its applicability depends on the boundary conditions and some other limiting conditions [197, 198]. As a correction, Brinkman proposed an extension in velocity [197, 199],

$$\mu \Delta \mathbf{v} - \mu/k\mathbf{v} = \nabla p, \quad \nabla \cdot \mathbf{v} = 0, \tag{11}$$

in which the Laplacian of the velocity field appears,[9] and this is called the Brinkman equation. That two terms in velocity distinguish two different length scales [197], and their transition occurs through the so-called Brinkman screening length, $\sqrt{k}$. According to the argument of Durlofsky and Brady [197], the Laplacian term dominates on small length scales, and it becomes negligible on large scales. The Brinkman model is investigated in detail by several authors [200–202].

It is experimentally shown that the porosity significantly influences the velocity field inside porous material, this is called 'channeling effect' [203–205]. We call the attention to the work of Vafai et al. [201, 206–209] in which numerous analytical and numerical solution methods are developed to describe the hydrodynamic and thermal behavior of porous materials. We also refer to the experimental work of Kowalski et al. [210–213] in which micro-scale diffusion processes are investigated.

---

[9] Analogously to the GK equation, it can be interpreted as a nonlocal term.

Biological Materials

One outstanding example of heterogeneous materials originate in biology. Their particular thermodynamic scaling properties are discussed in [214]. The organic structures are extremely complicated in many aspects. There is a continuous

- blood flow through the tissues,
- mass transfer between the tissues and blood,
- chemical reaction in the tissues,
- heat exchange on the interfaces,

and the variety of the structures is practically infinite. Consequently, there is no use a specific, structure-dependent model since even the same type of tissue differs between two people. Nevertheless, many thermal models in that field try to account the details, sometimes in an ad hoc way.

More importantly, there are parallel mechanisms in the heat transfer process. In a tissue, blood perfusion is present besides of metabolic heat generation, possible layered structure (e.g., in a skin) or heat exchange effect between artery-vein pairs. Essentially, these effects can be taken into account in the balance equation of internal energy. This is why so important to distinguish the balances from the constitutive relations. It does matter which one is extended. Although various transport mechanisms present, and it is possible to imagine a coupled model between the heat flux and the diffusion, for instance, the constitutive equations themselves cannot depend on the source terms. This is a general misleading aspect in the significant part of biological literature. Briefly, the most important biological models are related to Pennes [215], Chen and Holmes [216], Weinbaum, Jiji and Lemons [217–219], Wulff [220], Klinger [221, 222], Anderson and Valvano [223], Zolfaghari and Maerefat [224]. They differ in many aspects such as how they take into account the various physiological properties, such as the direction of the blood flow, body core temperature, vascular geometry, mechanical effects, environmental effects and so on [225–235]. Indeed, most of the models are constructed to keep the Fourier's law valid as much as possible; however, in some cases, the time derivatives are modified arbitrarily that cannot be done in that way due to objectivity reasons [236–238].

In Pennes' model, the blood perfusion and the metabolic heat generation effects are included as source terms,

$$\rho_t c_t \partial_t T = \lambda_t \partial_{xx} T + \rho_b c_b \omega_b (T - T_a) + q_{met}, \tag{12}$$

where ρ_t, c_t, λ_t are the tissue mass density, the specific heat, and the thermal conductivity, respectively, ω_b is the blood perfusion rate and ρ_b, c_b are the blood properties. Equation (12) describes the time evolution of tissue temperature, where the arterial blood temperature T_a is considered, and q_{met} being the metabolic heat generation. The blood perfusion is modeled as simple heat convection, but here the blood perfusion characterizes the heat transfer instead of a heat transfer coefficient. This is the simplest bio-heat conduction model which developed based on the steady temperature distribution of a forearm.

The Chen-Holmes (CH) and the Weinbaum-Jiji-Lemons (WJL) models are more advanced from a physiological point of view. They consist of several assumptions of specific biological properties such as the velocity field in the arteries and veins, even their arrangement is considered. These models are way too specific and cannot be applied in general situations. It is important to keep in mind that all these models obey the Fourier's law, and only source terms are introduced in the balance equation of internal energy. Wulff proposed the first non-Fourier biologically-motivated model:

$$\mathbf{q} = -\lambda_t \nabla T_t + \rho_b h_b \mathbf{U}_b, \tag{13}$$

where the indices t and b denote the tissue and blood properties as previously, h being the enthalpy and $\mathbf{U}_b$ is the local mean velocity of the blood. This model violates objectivity principles and does not compatible with the second law of thermodynamics. Moreover, the material parameters of blood and tissue are mixed, that makes this model questionable. Beyond the Wulff's model, usually, only the MCV and DPL equations are considered in most of the non-Fourier problems. It is worth to note that all of these models are very specific and hardly can be applied for general practical problems due to the lack of physiological information.

Banerjee et al. [239] performed a series of laser flash experiments with three different materials, namely processed meat, tissue phantoms, and fiber composites, with a macroscopic length scale. They successfully used the MCV equation to explain their results; nevertheless, there is no wave phenomenon in this experiment. In parallel, Jaunich et al. [240] did not manage to find MCV-like behavior in a similar flash experiment. Eventually, based on their results, the MCV equation is not appropriate for this task in contrary to Banerjee et al. The experiments made by Dhar et al. present temperature history not obeying the Fourier's law [241]. Then, not surprisingly, Pennes' model is also not applicable [241].

The DPL equation introduced by Tzou [242, 243] is popular in the field of biological heat conduction [244–246] despite its doubtful physical background [17, 247–253]. The fact that Taylor series expansion is not a thermodynamically consistent tool to derive a constitutive equation is ignored in a few studies [254, 255] and various versions of DPL are applied [244, 256, 257].

Despite the extensive criticism of the DPL equation, there are cases when it seems to be valid. For instance, a demonstrative analysis is made by Liu and Chen [258]. The DPL model is used for predicting the temperature distribution during a tumor treatment, and the solutions are compared to the experiments conducted by Andrä et al. [259]. Their results are in good agreement. On the contrary, Sahoo et al. investigated the temperature history in wet tissue and beside the DPL equation, they applied one of its generalized forms as well to model the laser irradiation [255]. One of the most important conclusion is that the mentioned Taylor series expansion is not only thermodynamically inconsistent but also introduces artificial effects in the solution that existence is not proved experimentally. Instead of the improvement of constitutive equations, it makes the model harder to solve and means a disadvantage from a theoretical point of view. Secondly, none of the DPL models were applicable.

The Guyer-Krumhansl equation also works successfully for materials with biological origins [17], thanks to the more general modeling level of NET-IV. In [17],
an experiment, performed by Tang et al. [260], is revisited. It turned out that the
size-dependence mentioned above observed in rock samples is present in biological
materials such as processed meat, too. However, it is necessary to point out that Tang
et al. have misinterpreted the measured over-diffusive propagation. They neglected
the possible cooling of the specimen. Therefore, the experiment is handled as an
adiabatic one. It leads to a completely different solution, and they have tried to fit the
DPL equation. It also turned out that in most cases, the Fourier equation is enough
applying the cooling boundary condition, and over-diffusive propagation is observed
only in one case [17].

Functionally Graded Materials

Previously, we have seen that in some particular case, a layered structure can lead to
non-Fourier behavior. It only occurs when the heat flux was parallel with the layers.
In analogy with the layered capacitor sample, we mention briefly the functionally
graded materials.

That type of structure consists of a transition from one material to the other which
can be continuous or step-wise [261]. Consequently, the material parameters are
spatial dependent, and their uncertainty around the transition can be significant as
well [262, 263]. Nevertheless, according to Fang and Hu [264], there are no direct
experimental results that would prove the presence of non-Fourier phenomenon.
However, it is not that clear as these phenomena depend on the length scales as well.
Recent results and analysis show a significant contribution of ballistic effects [113,
114, 265–267] and thermo-electric coupling [268–270].

An interesting theoretical analysis of FGM's is provided recently by Cao et
al. [271] in which a Guyer-Krumhansl-like model is proposed to model heat pulse
propagation. In contrary, Sladek et al. [272] applied the Fourier's law with space-
dependent coefficients. Here we would like to emphasize the direction dependence
both in the corresponding time and spatial scales and the possibility to find the deviation from the Fourier's law.

Nanofluids

A relatively new, pioneering field is related to nanofluids. In such a fluid—usually,
with low thermal conductivity—nano-sized particles with high thermal conductivity
are dispersed. It aims to improve the thermal properties of the fluid in order to
enhance heat transfer. Practically, the possible variations are unlimited, the number
of experiments is enormous. The literature that investigates the effective thermal
conductivity, the change in heat transfer coefficient and other thermal properties of
a nanofluid is immense. However, it is still not experimentally clear that whether a
nanofluid obeys the Fourier's law or not. Indeed, the behavior of a nanofluid depends

on many factors like the volume fraction of nanoparticles, flow states, temperature conditions, thermal properties, boundary conditions, and so on [273]. Although it makes the investigations more complicated, there is no conclusive report about the role of constitutive equations. There are many models to determine the effective thermal conductivity [274–281], analyzing the possible heat transfer mechanisms [277, 279, 280, 282–289], boiling properties [283] and a few studies that are dealing with non-Fourier constitutive equations [285, 287, 290, 291] but none of them have a unified conclusion in a sense that how to model a nanofluid transiently with given constituents. That is, there are no clear experimental results that could show clearly the deviation from Fourier's law. It does not mean that non-Fourier phenomena are not present in a nanofluid, but due to the various conditions and the limited possibilities of measurements, it could be much harder to find and obtain results like in case of low-temperature heat conduction. Moreover, Mohamad [292] also draws attention to some of the misleading theories related to the effective properties of nanofluids which strengthen the need for research on a more rigorous thermodynamical basis.

3   Scaling in Non-Newtonian Media

Previously, we focused our attention mostly on rigid solid materials in which some diffusion phenomenon occurs. Within this restricted situation, the time scales are given from the conductivity properties that influence the diffusion coefficient. For porous materials, the appearance of parallel time and length scales become visible. Its consequences have been measured both directly and indirectly. Now we move our focus on mechanics, briefly presenting the experimental evidence related to two particular fields. The first one is called rheology, where Hooke's law is not applicable. The second topic is about a more special problem, the modeling of rarefied gases.

3.1   Rheology

In the engineering practice, the usual assumption is that the body obeys the Hooke's law, i.e,

$$\sigma = E^{\text{dev}} \mathbf{D}^{\text{dev}} + E^{\text{sph}} \mathbf{D}^{\text{sph}}, \tag{14}$$

is applicable and for small deformation approximation, the elastic deformedness $\mathbf{D}$ reduces to the usual strain[10] ε. The coefficients $E^{\text{dev}} = 2G$ and $E^{\text{sph}} = 3K$, in

---

[10]The deformedness in a 1D case is defined as $D = \ln(l/l_r)$ with l_r being a relaxed length, characterizing the unloaded state of the material. This is an appropriate thermodynamic state variable. The engineering strain is $\varepsilon = (l - l_0)/l_0$ in which l_0 is the length at the initial time instant. Since ε is a reference time dependent, it is not a good thermodynamic state variable [293]. Unfortunately, D is not directly measurable.

Fig. 18 Thermal imaging during uniaxial loading [295]

which G and K are the shear and the bulk moduli, respectively, come from the Lamé coefficients. Their values, however, depend on the measurement: there is a significant difference between the static and dynamic loading (measured from the speed of sound) [294]. This is one important time scale that appears indirectly. Staying with purely elastic bodies, their thermal behavior during a simple uniaxial loading is essential, too. According to the related experiments [295, 296], the thermal effects are measurable and significant, it's modeling, however, is not trivial even in the elastic domain (Fig. 18).

To present the rheological time scales, we have to recall the Kluitenberg–Verhás model [297, 298] from Chap. 2, that can be decomposed into spherical and deviatoric parts in a three-dimensional situation,

$$\tau_{\text{dev}}\dot{\boldsymbol{\sigma}}_{\text{dev}} + \boldsymbol{\sigma}_{\text{dev}} = E_{0d}\mathbf{D}_{\text{dev}} + E_{1d}\dot{\mathbf{D}}_{\text{dev}} + E_{2d}\ddot{\mathbf{D}}_{\text{dev}}$$
$$\tau_{\text{sph}}\dot{\boldsymbol{\sigma}}_{\text{sph}} + \boldsymbol{\sigma}_{\text{sph}} = E_{0s}\mathbf{D}_{\text{sph}} + E_{1s}\dot{\mathbf{D}}_{\text{sph}} + E_{2s}\ddot{\mathbf{D}}_{\text{sph}}, \qquad (15)$$

where the upper dot simplifies our notation and denotes the time derivative and the indicies 'dev' and 'sph' refer to the deviatorical and spherical parts, respectively. Similarly to the Guyer-Krumhansl equation, it is possible to identify the new time scales in terms of τ, for both parts. We can continue the analogy, exploiting the hierarchy of the Hooke body in the Kluitenberg–Verhás model [294], hence introducing the 'Hooke resonance conditions',

$$b_d = \frac{E_{1d}}{\tau_d E_{0d}} \quad b_s = \frac{E_{1s}}{\tau_s E_{0s}}, \qquad (16)$$

where the ratios b_d and b_s are characterizing the deviation from the Hooke's law. According to the evaluation in [294], their values vary around $1.2-2$.

3.2 Rarefied Gases

Now we turn our attention to the limits of the Navier–Stokes–Fourier equations. Their generalizations are discussed in Chap. 2. To explain and predict the non-classical

Fig. 19 Arrangement of the
shock tube experiment [308].
With permission

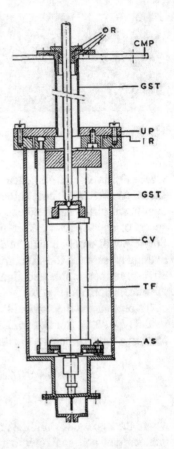

behavior, one needs to generalize not only the Fourier's law but the Navier–Stokes equation as well [299, 300]. In these models, the emphasis is on the coupling between the heat flux and the pressure, likewise in case of ballistic propagation. In this sense, the rarefied phonon gas is thermodynamically equivalent with a rarefied real gas [299], since the entropy itself does not make any distinction between different type of materials.

In the pioneering work of Meixner [301], only the fluid part is extended using the dynamic pressure as an internal variable [302], without direct coupling to the heat flux. Later on, Carrassi and Morro [303, 304], Jou et al. [305] and Ruggeri et al. [306, 307] developed their own rarefied gas models.

The corresponding experimental background is similar, but a bit different than the heat conduction case, it could be called as 'pressure pulse experiment' (shock tube experiment), see Fig. 19 for its arrangement.

On one side, the pressure field is perturbed, and the speed of sound is measured in the meantime. Beside the investigation of damping in such a rarefied fluid, the change in the speed of sound as a function of the mass density is recorded. One typical result is shown in Fig. 20. Usually, the density is decreased under 10 Pa, for instance, see the

Fig. 20 Typical
measurement results [312].
Later, the one that
corresponds to 296.8 K is
evaluated. With permission

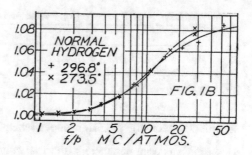

work of Meyer and Sessler [309]; besides, constant temperature was maintained. In
this state, the Knudsen number becomes large, goes above 0.05−0.1, and the Navier–
Stokes equation loses its validity [310, 311]. For a typical outcome, see Fig. 20.
Rhodes [312] performed that one, and other authors are obtained characteristically
similar results, such as Greenspan [313] and Sluijter et al. [308, 314].

There are only a few available evaluations of these experiments published recently.
An extensive review of these advancements is related to Arima et al. [306, 307, 315–
317] and Struchtrup [310, 311, 318–320]. In these works, many of the available
experimental data are evaluated. Here, we emphasize the scaling properties only, as
it appears in the experiments directly.

It is often stated that only the ratio of the frequency ω and the pressure p is
important, as it is also visible in Fig. 20. Consequently, the change in the speed
of sound is the function of ω/p only at a constant temperature. Indeed, such a
scaling property is the direct consequence of the Boltzmann equation in kinetic
theory. Since all experimental data are given as a function of ω/p, their evaluation
is straightforward using kinetic theory based models. However, from a continuum
point of view, the picture is not that clear.

For instance, let us restrict ourselves only the classical equations and calculate
the dispersion relations [299, 307] by assuming the common $e^{i(\omega t - kx)}$ plane wave
solution of the one-dimensional representation of the Navier–Stokes–Fourier system
(17), using the usual wave number k and frequency ω,

$$\partial_t \rho + \rho_0 \partial_x v = 0,$$
$$\rho_0 \partial_t v + \partial_x \Pi_d + \partial_x \Pi_s + RT_0 \partial_x \rho + R\rho_0 \partial_x T = 0,$$
$$\rho_0 c_V \partial_t T + \partial_x q + R\rho_0 T_0 \partial_x v = 0,$$
$$q + \lambda \partial_x T = 0,$$
$$\Pi_d + \nu \partial_x v = 0,$$
$$\Pi_s + \eta \partial_x v = 0, \tag{17}$$

where ν, η and λ are the shear, bulk viscosities and the thermal conductivity, respectively. Ideal gas state equation is already exploited. Then the following expression
for phase velocity $v_{ph} = \frac{\omega}{k}$ is obtained,

$$v_{ph}^2 = \frac{cRT\rho^2 + R^2T\rho^2 + ic\eta\rho\omega + i\lambda\rho\omega + ic\nu\rho\omega}{2c\rho^2} +$$

$$+ i\frac{\sqrt{\rho^2\left(-4c\lambda\omega(-iRT\rho + (\eta+\nu)\omega) + \left(iR^2T\rho - \lambda\omega + c(iRT\rho - (\eta+\nu)\omega)\right)^2\right)}}{2c\rho^2}.$$

$$(18)$$

In order to reconstruct the ω/p-dependence in Eq. (18), one must assume

- constant material parameters (thermal conductivity and viscosities), that is, they are independent of the mass density ρ,
- ideal gas state equation.

These constraints are apparent from continuum point of view and naturally embedded in the kinetic theory based models. The reason is that in kinetic theory, these transport coefficients are given directly since there is a detailed transport mechanism behind and all equations are derived accordingly. Although in a continuum model it is not possible to calculate the coefficients, but it is possible to measure them.

At that point, we highlight the main difference between a continuum and a kinetic model. While the kinetic approach fixes the coefficients as a function of temperature and pressure, the continuum one uses the measured values. This is the reason why one must distinguish the theoretical from the measured (or 'effective' or 'apparent') values.

In case of viscosity measurements of gases, both in rarefied (down to 0.1 Pa) and dense states (up to 1000 MPa), the density dependence is measured [321–327]. Despite the uncertainties, especially considering very low pressures, the density dependence is apparent. It is possible to embed that density dependence into a continuum theory as there is no any restriction in this respect. Then the resulting scaling would be contradictory, and the ω/p-dependence is not recovered. It holds for the generalized theories, too. However, it is not a problem if the frequency and the pressure are given separately, such as in the paper of Sette et al. [328].

From a kinetic point of view, it turned out that some corrections are necessary, both for rarefied and dense states. For instance, the Enskog-type correction [323, 329, 330] is in good agreement with dense state values. Michalis et al. draws the attention to the extensions for rarefied states, as a function of the Knudsen number [329, 331–334]. One more important consequence of the kinetic approach is about the zero-density limit of viscosity: it predicts non-zero viscosity for zero density. The corrections using the Knudsen number are able to overcome this contradiction [335, 336].

One particular experiment (see Fig. 21) is evaluated using the continuum model based on the approach of NET-IV [300]. In the generalized continuum model, the density dependence of the new parameters are considered in accordance with the kinetic theory: all of them is inversely proportional to the mass density; and the classical transport coefficients are constant: $\nu = 8.82 \cdot 10^{-6}$ Pas, $\eta = 326 \cdot 10^{-6}$ Pas and $\lambda = 0.182$ W/(mK). Tables 3 and 4 shows the fitted parameters [299]. The ratios of the relaxation times are adopted from the fitting of Arima et al. [307].

Fig. 21 Evaluation of
Rhodes' data with the RET
(red dashed line) and
NET-IV (thick black line)
models [299]. The pressure
starts at 1 atm and decreases
to 2000 Pa, $\omega = 1\,\text{MHz}$.
Error bars are placed for
each measurement point to
indicate the uncertainty of
digitalizing data; its
magnitude is $\pm 2.5\,\text{m/s}$. The
double horizontal scales
emphasize the equivalence
between the two scales as
only the mass density was
changed

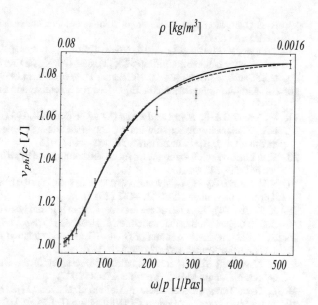

Table 3 Fitted relaxation time coefficients for continuum model

$\tau_q = \frac{t_1}{\rho}, t_1 = \left(\frac{\text{kg}\cdot\text{s}}{\text{m}^3}\right)$	$\tau_d = \frac{t_2}{\rho}, t_2 = \left(\frac{\text{kg}\cdot\text{s}}{\text{m}^3}\right)$	$\tau_s = \frac{t_3}{\rho}, t_3 = \left(\frac{\text{kg}\cdot\text{s}}{\text{m}^3}\right)$	τ_q/τ_d	τ_s/τ_d
$4.526 \cdot 10^{-9}$	$3.1 \cdot 10^{-9}$	$4.464 \cdot 10^{-7}$	1.46	144

Table 4 Fitted coupling coefficients for continuum model

$\alpha_{12} = \frac{a_{12}}{\rho}, a_{12} = \left(\frac{\text{kg}\cdot\text{s}}{\text{m}^3}\right)$	$\beta_{12} = \frac{b_{12}}{\rho}, b_{12} = \left(\frac{\text{kg}\cdot\text{s}}{\text{m}^3}\right)$	$\alpha_{21} = \frac{a_{21}}{\rho}, a_{21} = \left(\frac{\text{kg}}{\text{m}\cdot\text{s}}\right)$	$\beta_{21} = \frac{b_{21}}{\rho}, b_{21} = \left(\frac{\text{kg}}{\text{m}\cdot\text{s}}\right)$
$1.7 \cdot 10^{-6}$	$1.27 \cdot 10^{-6}$	$1.3 \cdot 10^{-4}$	$1.55 \cdot 10^{-5}$

References

1. J. Callaway, Model for lattice thermal conductivity at low temperatures. Phys. Rev. **113**(4), 1046 (1959)
2. Gy. Gróf, Notes on using temperature-dependent thermal diffusivity–forgotten rules. J. Therm. Anal. Calorim. **132**(2), 1389–1397 (2018)
3. S. Both, B. Czél, T. Fülöp, Gy. Gróf, Á. Gyenis, R. Kovács, P. Ván, J. Verhás, Deviation from the Fourier law in room-temperature heat pulse experiments. J. Non-Equilib. Thermodyn. **41**(1), 41–48 (2016)
4. P. Ván, A. Berezovski, T. Fülöp, Gy. Gróf, R. Kovács, Á. Lovas, J. Verhás, Guyer-Krumhansl-type heat conduction at room temperature. EPL, **118**(5), 50005 (2017), arXiv:1704.00341v1
5. W.J. Parker, R.J. Jenkins, C.P. Butler, G.L. Abbott, Flash method of determining thermal diffusivity, heat capacity, and thermal conductivity. J. Appl. Phys. **32**(9), 1679–1684 (1961)
6. H.M. James, Some extensions of the flash method of measuring thermal diffusivity. J. Appl. Phys. **51**(9), 4666–4672 (1980)

7. Gy.I. Gróf, Homogén és kétrétegű minták hőmérsékletvezetési tényezőjének mérése flash módszerrel (2002)
8. T. Fülöp, R. Kovács,Á Lovas, Á. Rieth, T. Fodor, M. Szücs, P. Ván, Gy. Gróf, Emergence of non-Fourier hierarchies. Entropy **20**(11), 832 (2018), ArXiv: 1808.06858
9. H.S. Carslaw, J.C. Jaeger, *Conduction of Heat in Solids* (1959)
10. S.J. Farlow, Partial Differential Equations for Scientists and Engineers (Courier Corporation, 1993)
11. M. Necati Ozisik, *Heat Conduction* (Wiley, New York, 1993)
12. K.V. Zhukovsky, Exact solution of Guyer-Krumhansl type heat equation by operational method. Int. J. Heat Mass Transf. **96**, 132–144 (2016)
13. K.V. Zhukovsky, Operational approach and solutions of hyperbolic heat conduction equations. Axioms **5**(4), 28 (2016)
14. K.V. Zhukovsky, H.M. Srivastava, Analytical solutions for heat diffusion beyond Fourier law. Appl. Math. Comput. **293**, 423–437 (2017)
15. K. Zhukovsky, Violation of the maximum principle and negative solutions for pulse propagation in Guyer-Krumhansl model. Int. J. Heat Mass Transf. **98**, 523–529 (2016)
16. R. Kovács, Analytic solution of Guyer-Krumhansl equation for laser flash experiments. Int. J. Heat Mass Transf. **127**, 631–636 (2018)
17. R. Kovács, P. Ván, Thermodynamical consistency of the Dual Phase Lag heat conduction equation. Contin. Mech. Thermodyn. 1–8 (2017)
18. L. Tisza, Transport phenomena in Helium II. Nature **141**, 913 (1938)
19. L. Landau, Two-fluid model of liquid Helium II. J. Phys. USSR **5**, 71 (1941)
20. L. Tisza, The theory of liquid Helium. Phys. Rev. **72**(9), 838–877 (1947)
21. L. Landau, On the theory of superfluidity of Helium II. J. Phys. **11**(1), 91–92 (1947)
22. S. Balibar, Laszlo Tisza and the two-fluid model of superfluidity. Comptes Rendus Phys. **18**(9–10), 586–591 (2017)
23. V. Peshkov, Second sound in Helium II. J. Phys. (Moscow) **381**(8) (1944)
24. P.L. Kapitza, Heat transfer and superfluidity of helium II. Phys. Rev. **60**(4), 354 (1941)
25. F. London, *Superfluids*, Structure of Matter Series (Wiley, New York, 1954)
26. R.J. Donnelly, The two-fluid theory and second sound in liquid Helium. Phys. Today **62**(10), 34–39 (2009)
27. P.C. Hohenberg, P.C. Martin, Microscopic theory of superfluid helium. Ann. Phys. **34**(2), 291–359 (1965)
28. P.W. Anderson, Considerations on the flow of superfluid helium. Rev. Mod. Phys. **38**(2), 298 (1966)
29. S. J. Putterman. Superfluid hydrodynamics. In *Amsterdam, North-Holland Publishing Co.; New York, American Elsevier Publishing Co., Inc.(North-Holland Series in Low Temperature Physics. Volume 3), 1974. 464 p.*, volume 3, 1974
30. G.P. Bewley, D.P. Lathrop, K.R. Sreenivasan, Superfluid helium: visualization of quantized vortices. Nature **441**(7093), 588 (2006)
31. L. Dresner, *Transient Heat Transfer in Superfluid Helium*, vol. 27 (Plenum Press, New York, 1982)
32. L. Dresner, *Transient Heat Transfer in Superfluid Helium- Part II* (Springer, Berlin, 1984)
33. E. Kim, M.H.W. Chan, Probable observation of a supersolid helium phase. Nature **427**(6971), 225 (2004)
34. D. Vollhardt, P. Wölfle, *The Superfluid Phases of Helium 3* (Courier Corporation, 2013)
35. M.S. Mongiovi, D. Jou, M. Sciacca, Non-equilibrium thermodynamics, heat transport and thermal waves in laminar and turbulent superfluid helium. Phys. Rep. (2018)
36. J.R. Pellam, Investigations of pulsed second sound in liquid helium II. Phys. Rev. **75**(8), 1183 (1949)
37. V. Narayanamurti, R.C. Dynes, K. Andres, Propagation of sound and second sound using heat pulses. Phys. Rev. B **11**(7), 2500–2524 (1975)
38. C.T. Lane, H. Fairbank, H. Schultz, W. Fairbank, "Second sound" in liquid Helium II. Phys. Rev. **70**(5–6), 431 (1946)

39. C.T. Lane, H.A. Fairbank, W.M. Fairbank, Second sound in liquid Helium II. Phys. Rev. **71**, 600–605 (1947)
40. R.D. Maurer, M.A. Herlin, Second sound velocity in Helium II. Phys. Rev. **76**(7), 948 (1949)
41. J.C. Ward, J. Wilks, The velocity of second sound in liquid Helium near the absolute zero. Lond. Edinb. Dublin Philos. Mag. J. Sci. **42**(326), 314–316 (1951)
42. K.R. Atkins, D.V. Osborne, The velocity of second sound below 1 K. Lond. Edinb. Dublin Philos. Mag. J. Sci. **41**(321), 1078–1081 (1950)
43. M. Chester, Second sound in solids. Phys. Rev. **131**(5), 2013–2015 (1963)
44. R.A. Guyer, J.A. Krumhansl, Dispersion relation for second sound in solids. Phys. Rev. **133**(5A), A1411 (1964)
45. C.C. Ackerman, B. Bertman, H.A. Fairbank, R.A. Guyer, Second sound in solid Helium. Phys. Rev. Lett. **16**(18), 789–791 (1966)
46. R.J. Hardy, S.S. Jaswal, Velocity of second sound in NaF. Phys. Rev. B **3**(12), 4385–4387 (1971)
47. V. Narayanamurti, R.C. Dynes, Observation of second sound in bismuth. Phys. Rev. Lett. **28**(22), 1461–1465 (1972)
48. H.A. Fairbank, K.H. Mueller, Propagation of second sound and heat pulses in solid helium crystals, *Quantum Statistical Mechanics in the Natural Sciences* (Springer, Berlin, 1974), pp. 403–411
49. A. Sellitto, V.A. Cimmelli, D. Jou, *Mesoscopic Theories of Heat Transport in Nanosystems*, vol. 6 (Springer, Berlin, 2016)
50. Y. Guo, D. Jou, M. Wang, Macroscopic heat transport equations and heat waves in nonequilibrium states. Phys. D: Nonlinear Phenom. **342**, 24–31 (2017)
51. D.D. Joseph, L. Preziosi, Heat waves. Rev. Mod. Phys. **61**(1), 41 (1989)
52. D.D. Joseph, L. Preziosi, Addendum to the paper on heat waves. Rev. Mod. Phys. **62**(2), 375–391 (1990)
53. P. Ván, Theories and heat pulse experiments of non-Fourier heat conduction. Commun. Appl. Ind. Math. **7**(2), 150–166 (2016)
54. V.A. Cimmelli, Different thermodynamic theories and different heat conduction laws. J. Non-Equilib. Thermodyn. **34**(4), 299–333 (2009)
55. V.A. Cimmelli, A. Sellitto, D. Jou, Nonlocal effects and second sound in a non-equilibrium steady state. Phys. Rev. B **79**(1), 014303 (2009)
56. D. Jou, V.A. Cimmelli, Constitutive equations for heat conduction in nanosystems and non-equilibrium processes: an overview. Commun. Appl. Ind. Math. **7**(2), 196–222 (2016)
57. D. Jou, I. Carlomagno, V.A. Cimmelli, A thermodynamic model for heat transport and thermal wave propagation in graded systems. Phys. E: Low-dimens. Syst. Nanostruct. **73**, 242–249 (2015)
58. A. Sellitto, V.A. Cimmelli, D. Jou, Nonequilibrium thermodynamics and heat transport at nanoscale, *Mesoscopic Theories of Heat Transport in Nanosystems* (Springer International Publishing, Berlin, 2016), pp. 1–30
59. I. Carlomagno, A. Sellitto, V.A. Cimmelli, Dynamical temperature and generalized heat-conduction equation. Int. J. Non-Linear Mech. **79**, 76–82 (2016)
60. G. Mascali, V. Romano, Charge transport in graphene including thermal effects. SIAM J. Appl. Math. **77**(2), 593–613 (2017)
61. H.E. Jackson, C.T. Walker, T.F. McNelly, Second sound in NaF. Phys. Rev. Lett. **25**(1), 26–28 (1970)
62. T.F. McNelly, S.J. Rogers, D.J. Channin, R.J. Rollefson, W.M. Goubau, G.E. Schmidt, J.A. Krumhansl, R.O. Pohl, Heat pulses in NaF: onset of second sound. Phys. Rev. Lett. **24**(3), 100–102 (1970)
63. H.E. Jackson, C.T. Walker, Thermal conductivity, second sound and phonon-phonon interactions in NaF. Phys. Rev. B **3**(4), 1428–1439 (1971)
64. T.F. McNelly, Second Sound and Anharmonic Processes in Isotopically Pure Alkali-Halides, Ph.D. Thesis, Cornell University (1974)

65. C.T. Walker, Thermal conductivity of some alkali halides containing F centers. Phys. Rev. **132**(5), 1963–1975 (1963)
66. L.P. Mezhov-Deglin, Measurement of the thermal conductivity of crystalline he4. Sov. Phys. JETP **22**, 47 (1966)
67. P.D. Thacher, Effect of boundaries and isotopes on the thermal conductivity of LiF. Phys. Rev. **156**(3), 975 (1967)
68. S.B. Trickey, W.P. Kirk, E.D. Adams, Thermodynamic, elastic, and magnetic properties of solid helium. Rev. Mod. Phys. **44**(4), 668 (1972)
69. R.H. Crepeau, O. Heybey, D.M. Lee, S.A. Strauss, Sound propagation in hcp solid helium crystals of known orientation. Phys. Rev. A **3**(3), 1162 (1971)
70. P.V.E. McClintock, An apparatus for preparing isotopically pure he4. Cryogenics **18**(4), 201–208 (1978)
71. R.J. Von Gutfeld, Heat pulse transmission. Phys. Acoust. **5**, 233 (2012)
72. S.J. Rogers, Transport of heat and approach to second sound in some isotropically pure alkali-halide crystals. Phys. Rev. B **3**(4), 1440 (1971)
73. S.J. Rogers, Second sound in solids: the effects of collinear and non-collinear three phonon processes. Le Journal de Physique Colloques **33**(4), 4–111 (1972)
74. W. Dreyer, H. Struchtrup, Heat pulse experiments revisited. Contin. Mech. Thermodyn. **5**, 3–50 (1993)
75. K. Frischmuth, V.A. Cimmelli, Numerical reconstruction of heat pulse experiments. Int. J. Eng. Sci. **33**(2), 209–215 (1995)
76. K. Frischmuth, V.A. Cimmelli, Hyperbolic heat conduction with variable relaxation time. J. Theor. Appl. Mech. **34**(1), 57–65 (1996)
77. K. Frischmuth, V.A. Cimmelli, Coupling in thermo-mechanical wave propagation in NaF at low temperature. Arch. Mech. **50**(4), 703–713 (1998)
78. G. Chen, Ballistic-diffusive heat-conduction equations. Phys. Rev. Lett. **86**(11), 2297–2300 (2001)
79. G. Chen, Ballistic-diffusive equations for transient heat conduction from nano to macroscales. J. Heat Transf. **124**(2), 320–328 (2002)
80. Y. Ma, A hybrid phonon gas model for transient ballistic-diffusive heat transport. J. Heat Transf. **135**(4), 044501 (2013)
81. Y. Ma, A transient ballistic– diffusive heat conduction model for heat pulse propagation in nonmetallic crystals. Int. J. Heat Mass Transf. **66**, 592–602 (2013)
82. Y. Ma, Equation of phonon hydrodynamics for non-Fourier heat conduction, in *44th AIAA Thermophysics Conference*, pp. 2902 (2013)
83. F.X. Alvarez, D. Jou, Memory and nonlocal effects in heat transport: from diffusive to ballistic regimes. Appl. Phys. Lett. **90**(8), 083109 (2007)
84. D. Jou, J. Casas-Vázquez, G. Lebon, Extended irreversible thermodynamics. Rep. Prog. Phys. **51**(8), 1105 (1988)
85. D. Jou, J. Casas-Vázquez, G. Lebon, M. Grmela, A phenomenological scaling approach for heat transport in nano-systems. Appl. Math. Lett. **18**(8), 963–967 (2005)
86. G. Lebon, M. Grmela, C. Dubois, From ballistic to diffusive regimes in heat transport at nano-scales. Comptes Rendus Mecanique **339**(5), 324–328 (2011)
87. G. Lebon, M. Hatim, M. Grmela, Ch. Dubois, An extended thermodynamic model of transient heat conduction at sub-continuum scales. **467**(2135), 3241–3256 (2011)
88. D. Jou, J. Casas-Vazquez, G. Lebon, Extended irreversible thermodynamics revisited (1988–98). Rep. Prog. Phys. **62**(7), 1035 (1999)
89. F.X. Alvarez, D. Jou, Boundary conditions and evolution of ballistic heat transport. J. Heat Transf. **132**(1), 012404 (2010)
90. R. Kovács, P. Ván, Generalized heat conduction in heat pulse experiments. Int. J. Heat Mass Transf. **83**, 613–620 (2015)
91. R. Kovács, P. Ván, Models of ballistic propagation of heat at low temperatures. Int. J. Thermophys. **37**(9), 95 (2016)

92. R. Kovács, P. Ván, Second sound and ballistic heat conduction: NaF experiments revisited. Int. J. Heat Mass Transf. **117**, 682–690 (2018), arXiv:1708.09770

93. L.D. Landau, E.M. Lifshitz, *Theoretical Physics*, vol. 6 (Fluid Mechanics. Nauka, Moscow, 1986)

94. A. Berezovski, M. Berezovski, Influence of microstructure on thermoelastic wave propagation. Acta Mechanica **224**(11), 2623–2633 (2013)

95. A. Berezovski, J. Engelbrecht, P. Ván, Weakly nonlocal thermoelasticity for microstructured solids: microdeformation and microtemperature. Arch. Appl. Mech. **84**(9–11), 1249–1261 (2014)

96. A. Berezovski, P. Ván, Microdeformation and microtemperature. pp. 175–190 (2017)

97. A. Berezovski, P. Ván, *Internal Variables in Thermoelasticity* (Springer, Berlin, 2017)

98. P. Ván, A. Berezovski, J. Engelbrecht, Internal variables and dynamic degrees of freedom. J. Non-Equilib. Thermodyn. **33**(3), 235–254 (2008)

99. A. Berezovski, J. Engelbrecht, G.A. Maugin, Thermoelasticity with dual internal variables. J. Therm. Stresses **34**(5–6), 413–430 (2011)

100. I. Müller, T. Ruggeri, *Rational Extended Thermodynamics* (Springer, Berlin, 1998)

101. B. Nyíri, On the entropy current. J. Non-Equilib. Thermodyn. **16**(2), 179–186 (1991)

102. J. Verhás, *Thermodynamics and Rheology* (Akadémiai Kiadó-Kluwer Academic Publisher, 1997)

103. A. Berezovski and P. Ván. Internal variables in thermoelasticity. In Gy. Gróf and R. Kovács, editors, *MS Abstract book of the 14th Joint European Thermodynamics Conference*, pages 102–104, Budapest, 2017. Department of Energy Engineering, BME. ISBN 978-963-313-259-3

104. R. Kovács, *Heat conduction beyond Fourier's law: theoretical predictions and experimental validation*. Ph.D. thesis, Budapest University of Technology and Economics (BME) (2017)

105. B.D. Coleman, D.C. Newman, Implications of a nonlinearity in the theory of second sound in solids. Phys. Rev. B **37**(4), 1492 (1988)

106. E. Parthãe, L. Gmelin, *Gmelin Handbook of Inorganic and Organometallic Chemistry: TYPIX. Standardized Data and Crystal Chemical Characterization of Inorganic Structure Types*, vol. 2 (Springer, Berlin, 1993)

107. S. Bargmann, P. Steinmann, Finite element approaches to non-classical heat conduction in solids. Comput. Model. Eng. Sci. **9**(2), 133–150 (2005)

108. P.G. Klemens, Theory of thermal conduction in thin ceramic films. Int. J. Thermophys. **22**(1), 265–275 (2001)

109. D.G. Cahill, K. Goodson, A. Majumdar, Thermometry and thermal transport in micro/nanoscale solid-state devices and structures. J. Heat Transf. **124**(2), 223–241 (2002)

110. D.G. Cahill, W.K. Ford, K.E. Goodson, G.D. Mahan, A. Majumdar, H.J. Maris, R. Merlin, S.R. Phillpot, Nanoscale thermal transport. J. Appl. Phys. **93**(2), 793–818 (2003)

111. W. Kim, R. Wang, A. Majumdar, Nanostructuring expands thermal limits. Nano Today **2**(1), 40–47 (2007)

112. V. Rawat, Y.K. Koh, D.G. Cahill, T.D. Sands, Thermal conductivity of (Zr, W) N/ScN metal/semiconductor multilayers and superlattices. J. Appl. Phys. **105**(2), 024909 (2009)

113. B. Saha, T.D. Sands, U.V. Waghmare, First-principles analysis of ZrN/ScN metal/semiconductor superlattices for thermoelectric energy conversion. J. Appl. Phys. **109**(8), 083717 (2011)

114. B. Saha, Y.R. Koh, J. Comparan, S. Sadasivam, J.L. Schroeder, M. Garbrecht, A. Mohammed, J. Birch, T. Fisher, A. Shakouri, T.D. Sands, Cross-plane thermal conductivity of (Ti, W) N/(Al, Sc) N metal/semiconductor superlattices. Phys. Rev. B **93**(4), 045311 (2016)

115. W. Liu, M. Asheghi, Phonon-boundary scattering in ultrathin single-crystal silicon layers. Appl. Phys. Lett. **84**(19), 3819–3821 (2004)

116. Z. Hao, L. Zhichao, T. Lilin, T. Zhimin, L. Litian, L. Zhijian, Measurement of thermal conductivity of ultra-thin single crystal silicon film using symmetric structure. Chin. J. Semiconductors-Chin. Ed. **27**(11), 1961 (2006)

117. F. Vázquez, F. Márkus, K. Gambár, Quantized heat transport in small systems: a phenomenological approach. Phys. Rev. E **79**(3), 031113 (2009)
118. F. Vázquez, F. Márkus, Size effects on heat transport in small systems: dynamical phase transition from diffusive to ballistic regime. J. Appl. Phys. **105**(6), 064915 (2009)
119. M. Wang, N. Yang, Z.-Y. Guo, Non-Fourier heat conductions in nanomaterials. J. Appl. Phys. **110**(6), 064310 (2011)
120. A.I. Hochbaum, R. Chen, R.D. Delgado, W. Liang, E.C. Garnett, M. Najarian, A. Majumdar, P. Yang, Enhanced thermoelectric performance of rough silicon nanowires. Nature **451**(7175), 163 (2008)
121. S.R. Choi, D. Kim, S.-H. Choa, S.-H. Lee, J.-K. Kim, Thermal conductivity of AlN and SiC thin films. Int. J. Thermophys. **27**(3), 896–905 (2006)
122. N. Yang, G. Zhang, B. Li, Violation of Fourier's law and anomalous heat diffusion in silicon nanowires. Nano Today **5**(2), 85–90 (2010)
123. R. Chen, A.I. Hochbaum, P. Murphy, J. Moore, P. Yang, A. Majumdar, Thermal conductance of thin silicon nanowires. Phys. Rev. Lett. **101**(10), 105501 (2008)
124. M. Fujii, X. Zhang, H. Xie, H. Ago, K. Takahashi, T. Ikuta, Hi. Abe, T. Shimizu, Measuring the thermal conductivity of a single carbon nanotube. Phys. Rev. Lett. **95**(6), 065502 (2005)
125. B.-Y. Cao, Z.-Y. Guo, Equation of motion of a phonon gas and non-Fourier heat conduction. J. Appl. Phys. **102**(5), 053503 (2007)
126. P.T. Alvarez, *Thermal Transport in Semiconductors* (Springer, Berlin, 2017)
127. A. Ziabari, P. Torres, B. Vermeersch, Y. Xuan, X. Cartoixà, A. Torelló, J.-H. Bahk, Y.R. Koh, M. Parsa, D.Y. Peide, F.X. Alvarez, A. Shakouri, Full-field thermal imaging of quasiballistic crosstalk reduction in nanoscale devices. Nat. Commun. **9**(1), 255 (2018)
128. J. Shiomi, S. Maruyama, Non-Fourier heat conduction in a single-walled carbon nanotube: classical molecular dynamics simulations. Phys. Rev. B **73**(20), 205420 (2006)
129. C.-W. Chang, D. Okawa, H. Garcia, A. Majumdar, A. Zettl, Breakdown of Fourier's law in nanotube thermal conductors. Phys. Rev. Lett. **101**(7), 075903 (2008)
130. A. Cepellotti, G. Fugallo, L. Paulatto, M. Lazzeri, F. Mauri, N. Marzari, Phonon hydrodynamics in two-dimensional materials. Nat. Commun. **6**, 6400 (2015)
131. J.L. Vossen, W. Kern, W. Kern, *Thin Film Processes II*, vol. 2 (Gulf Professional Publishing, Houston, 1991)
132. M. Ohring, *Materials Science of Thin Films* (Elsevier, New York, 2001)
133. A. Majumdar, Microscale heat conduction in dielectric thin films. J. Heat Transf. **115**(1), 7–16 (1993)
134. A.A. Joshi, A. Majumdar, Transient ballistic and diffusive phonon heat transport in thin films. J. Appl. Phys. **74**(1), 31–39 (1993)
135. G. Chen, Phonon wave heat conduction in thin films and superlattices. J. Heat Transf. **121**(4), 945–953 (1999)
136. G. Chen, Particularities of heat conduction in nanostructures. J. Nanoparticle Res. **2**(2), 199–204 (2000)
137. G. Chen, Phonon heat conduction in nanostructures. Int. J. Therm. Sci. **39**(4), 471–480 (2000)
138. S.D. Brorson, J.G. Fujimoto, E.P. Ippen, Femtosecond electronic heat-transport dynamics in thin gold films. Phys. Rev. Lett. **59**(17), 1962 (1987)
139. S.D. Brorson, A. Kazeroonian, J.S. Moodera, D.W. Face, T.K. Cheng, E.P. Ippen, M.S. Dresselhaus, G. Dresselhaus, Femtosecond room-temperature measurement of the electron-phonon coupling constant γ in metallic superconductors. Phys. Rev. Lett. **64**(18), 2172 (1990)
140. J. Hohlfeld, J.G. Müller, S.-S. Wellershoff, E. Matthias, Time-resolved thermoreflectivity of thin gold films and its dependence on film thickness. Appl. Phys. B **64**(3), 387–390 (1997)
141. M.E. Siemens, Q. Li, R. Yang, K.A. Nelson, E.H. Anderson, M.M. Murnane, H.C. Kapteyn, Quasi-ballistic thermal transport from nanoscale interfaces observed using ultrafast coherent soft X-ray beams. Nat. Mater. **9**(1), 26 (2010)
142. K.M. Hoogeboom-Pot, J.N. Hernandez-Charpak, X. Gu, T.D. Frazer, E.H. Anderson, W. Chao, R.W. Falcone, R. Yang, M.M. Murnane, H.C. Kapteyn, D. Nardi. A new regime of nanoscale thermal transport: Collective diffusion increases dissipation efficiency, in *Proceedings of the National Academy of Sciences* (2015), pp. 201503449

143. J. Lee, J. Lim, P. Yang, Ballistic phonon transport in holey silicon. Nano Lett. **15**(5), 3273–3279 (2015)
144. C.Y. Zhao, T.J. Lu, H.P. Hodson, J.D. Jackson, The temperature dependence of effective thermal conductivity of open-celled steel alloy foams. Mater. Sci. Eng.: A **367**(1–2), 123–131 (2004)
145. F.A. Coutelieris, J.M.P.Q. Delgado, *Transport Processes in Porous Media* (Springer, Berlin, 2012)
146. K. Bamdad, A. Azimi, H. Ahmadikia, Thermal performance analysis of arbitrary-profile fins with non-fourier heat conduction behavior. J. Eng. Math. **76**(1), 181–193 (2012)
147. T.J. Bright, Z.M. Zhang, Common misperceptions of the hyperbolic heat equation. J. Thermophys. Heat Transf. **23**, 601–607 (2009)
148. R. Singh, H.S. Kasana, Computational aspects of effective thermal conductivity of highly porous metal foams. Appl. Therm. Eng. **24**(13), 1841–1849 (2004)
149. M.A. Schuetz, L.R. Glicksman, A basic study of heat transfer through foam insulation. J. Cell. Plast. **20**(2), 114–121 (1984)
150. J.M.P.Q. Delgado, *Heat and Mass Transfer in Porous Media* (Springer, Berlin, 2012)
151. A. Bhattacharya, V.V. Calmidi, R.L. Mahajan, Thermophysical properties of high porosity metal foams. Int. J. Heat Mass Transf. **45**(5), 1017–1031 (2002)
152. A.M. Druma, M.K. Alam, C. Druma, Analysis of thermal conduction in carbon foams. Int. J. Therm. Sci. **43**(7), 689–695 (2004)
153. F.A.L. Dullien, *Porous Media: Fluid Transport and Pore Structure* (Academic, 2012)
154. Z. Liu, *Multiphysics in Porous Materials* (Springer International Publishing AG, 2018)
155. A.G. Leach, The thermal conductivity of foams I: models for heat conduction. J. Phys. D: Appl. Phys. **26**(5), 733 (1993)
156. A.V. Luikov, Application of the methods of thermodynamics of irreversible processes to the investigation of heat and mass transfer. J. Eng. Phys. **9**(3), 189–202 (1965)
157. A.V. Luikov, Application of irreversible thermodynamics methods to investigation of heat and mass transfer. Int. J. Heat Mass Transf. **9**(2), 139–152 (1966)
158. W. Kaminski, Hyperbolic heat conduction equation for materials with a nonhomogeneous inner structure. J. Heat Transf. **112**(3), 555–560 (1990)
159. A. Graßmann, F. Peters, Experimental investigation of heat conduction in wet sand. Heat Mass Transf. **35**(4), 289–294 (1999)
160. H. Herwig, K. Beckert, Fourier versus non-Fourier heat conduction in materials with a non-homogeneous inner structure. Trans.-Am. Soc. Mech. Eng. J. Heat Transf. **122**(2), 363–364 (2000)
161. H. Herwig, K. Beckert, Experimental evidence about the controversy concerning Fourier or non-Fourier heat conduction in materials with a nonhomogeneous inner structure. Heat Mass Transf. **36**(5), 387–392 (2000)
162. A. Vedavarz, S. Kumar, M.K. Moallemi, Significance of non-Fourier heat waves in conduction. J. Heat Transf. **116**(1), 221–226 (1994)
163. K. Mitra, S. Kumar, A. Vedevarz, M.K. Moallemi, Experimental evidence of hyperbolic heat conduction in processed meat. J. Heat Transf. **117**(3), 568–573 (1995)
164. E.P. Scott, M. Tilahun, B. Vick, The question of thermal waves in heterogeneous and biological materials. J. Biomech. Eng. **131**(7), 074518 (2009)
165. W. Roetzel, N. Putra, S.K. Das, Experiment and analysis for non-Fourier conduction in materials with non-homogeneous inner structure. Int. J. Therm. Sci. **42**(6), 541–552 (2003)
166. P.J. Antaki, New interpretation of non-Fourier heat conduction in processed meat. J. Heat Transf. **127**(2), 189–193 (2005)
167. F. Jiang, Non-Fourier heat conduction phenomena in porous material heated by microsecond laser pulse. Microscale Thermophys. Eng. **6**(4), 331–346 (2003)
168. D.S. Chandrasekharaiah, Hyperbolic thermoelasticity: a review of recent literature. Appl. Mech. Rev. **51**(12), 705–729 (1998)
169. R.E. Khayat, J. deBruyn, M. Niknami, D.F. Stranges, R.M.H. Khorasany, Non-Fourier effects in macro-and micro-scale non-isothermal flow of liquids and gases. review. Int. J. Therm. Sci. **97**, 163–177 (2015)

170. L. Kovács, E. Mészáros, F. Deák, G. Somodi, K. Máté, A. Jakab (Kőmérő Kft.), Vásárhelyi
     B. (Vásárhelyi és Társa Kft.), Geiger J. (SZTE), Dankó Gy., Korpai F., Mező Gy., Darvas
     K. (Golder Zrt.), Ván P., Fülöp T., and Asszonyi Cs. (Montavid Termodinamikai Kutatóc-
     soport). A Geotechnikai Értelmező Jelentés (GÉJ) felülvizsgálata és kiterjesztése. Technical
     report. Kézirat - Kőmérő Kft. Pécs, RHK Kft. Irattár, RHK-K-032/12 (2012)
171. B. Czél, T. Fülöp, Gy. Gróf, Á. Gyenis, P. Ván, Simple heat conduction experiments, in *11th
     International Conference on Heat Engines and Environmental Protection*, ed. by Dombi Sz,
     Budapest, BME, Dep. of Energy Engineering (2013), pp. 141–146
172. T. Fülöp, R. Kovács, P. Ván, Thermodynamic hierarchies of evolution equations. Proc. Esto-
     nian Acad. Sci. **64**(3), 389–395 (2015)
173. A. Kossa, A new biaxial compression fixture for polymeric foams. Polym. Test. **45**, 47–51
     (2015)
174. A. Kossa, Sz. Berezvai, Visco-hyperelastic characterization of polymeric foam materials.
     Mater. Today: Proc. **3**(4), 1003–1008 (2016)
175. M.F. Ashby, T. Evans, N.A. Fleck, J.W. Hutchinson, H.N.G. Wadley, L.J. Gibson, *Metal
     Foams: A Design Guide* (Elsevier, 2000)
176. D.L. Weaire, S. Hutzler, *The Physics of Foams* (Oxford University Press, Oxford, 2001)
177. M.A. Mujeebu, M. Zu. Abdullah, M.Z.A. Bakar, A.A. Mohamad, M.K. Abdullah, Appli-
     cations of porous media combustion technology–a review. Appl. Energy **86**(9), 1365–1375
     (2009)
178. M.A. Mujeebu, M.Z. Abdullah, A.A. Mohamad, M.A. Bakar, Trends in modeling of porous
     media combustion. Prog. Energy Combust. Sci. **36**(6), 627–650 (2010)
179. D. Trimis, F. Durst, Combustion in a porous medium-advances and applications. Combust.
     Sci. Technol. **121**(1–6), 153–168 (1996)
180. N.I. Kim, S. Kato, T. Kataoka, T. Yokomori, S. Maruyama, T. Fujimori, K. Maruta, Flame
     stabilization and emission of small Swiss-roll combustors as heaters. Combust. Flame **141**(3),
     229–240 (2005)
181. N.I. Kim, S. Aizumi, T. Yokomori, S. Kato, T. Fujimori, K. Maruta, Development and scale
     effects of small swiss-roll combustors. Proc. Combust. Inst. **31**(2), 3243–3250 (2007)
182. S.K. Som, A. Datta, Thermodynamic irreversibilities and exergy balance in combustion pro-
     cesses. Prog. Energy Combust. Sci. **34**(3), 351–376 (2008)
183. S. Wood, A.T. Harris, Porous burners for lean-burn applications. Prog. Energy Combust. Sci.
     34(5), 667–684 (2008)
184. Y. Ju, K. Maruta, Microscale combustion: technology development and fundamental research.
     Prog. Energy Combust. Sci. **37**(6), 669–715 (2011)
185. S. Whitaker, Simultaneous heat, mass, and momentum transfer in porous media: a theory of
     drying. *Advances in Heat Transfer*, vol. 13 (Elsevier, 1977). pp. 119–203
186. L. Imre, T. Környey, Computer simulation of salami drying. Int. J. Numer. Methods Eng.
     30(4), 767–777 (1990)
187. T.Z. Harmathy, Simultaneous moisture and heat transfer in porous systems with particular
     reference to drying. Ind. Eng. Chem. Fundam. **8**(1), 92–103 (1969)
188. I. Farkas, M.J. Lampinen, K. Ojala, Water flow and binder migration during drying of coated
     paper. Dry. Technol. **9**(4), 1019–1049 (1991)
189. C.L.D. Huang, H.H. Siang, C.H. Best, Heat and moisture transfer in concrete slabs. Int. J.
     Heat Mass Transf. **22**(2), 257–266 (1979)
190. S. Whitaker, flow in porous media i: a theoretical derivation of darcy's law. Transp. Porous
     Media **1**(1), 3–25 (1986)
191. M. Liu, J. Wu, Y. Gan, D.A.H. Hanaor, C.Q. Chen, Evaporation limited radial capillary
     penetration in porous media. Langmuir **32**(38), 9899–9904 (2016)
192. G. Rehage, O. Ernst, J. Fuhrmann, Fickian and non-Fickian diffusion in high polymer systems.
     Discuss. Faraday Soc. **49**, 208–221 (1970)
193. D. Jou, J. Casas-Vázquez, G. Lebon, *Extended Irreversible Thermodynamics*, 4th edn.
     (Springer, New York, 2010)

194. D. Jou, J. Camacho, M. Grmela, On the nonequilibrium thermodynamics of non-Fickian diffusion. Macromolecules **24**(12), 3597–3602 (1991)
195. E.H. Wong, K.C. Chan, T.B. Lim, T.F. Lam, Non-Fickian moisture properties characterisation and diffusion modeling for electronic packages (1999), pp. 302–306
196. D. De Kee, Q. Liu, J. Hinestroza, Viscoelastic (non-Fickian) diffusion. Canad. J. Chem. Eng. **83**(6), 913–929 (2005)
197. L. Durlofsky, J.F. Brady, Analysis of the Brinkman equation as a model for flow in porous media. Phys. Fluids **30**(11), 3329–3341 (1987)
198. K.Y. Wertheim, T. Roose, A mathematical model of lymphangiogenesis in a zebrafish embryo. Bull. Math. Biol. **79**(4), 693–737 (2017)
199. H.C. Brinkman, A calculation of the viscous force exerted by a flowing fluid on a dense swarm of particles. Flow, Turbul. Combust. **1**(1), 27 (1949)
200. D.A. Nield, The limitations of the Brinkman-Forchheimer equation in modeling flow in a saturated porous medium and at an interface. Int. J. Heat Fluid Flow **12**(3), 269–272 (1991)
201. K. Vafai, S. Kim, On the limitations of the Brinkman-Forchheimer-extended Darcy equation. Int. J.f Heat Fluid Flow **16**(1), 11–15 (1995)
202. F.J. Valdes-Parada, J.A. Ochoa-Tapia, J. Alvarez-Ramirez, On the effective viscosity for the darcy-brinkman equation. Phys. A: Stat. Mech. Appl. **385**(1), 69–79 (2007)
203. K. Vafai, Convective flow and heat transfer in variable-porosity media. J. Fluid Mech. **147**, 233–259 (1984)
204. K. Vafai, R.L. Alkire, C.L. Tien, An experimental investigation of heat transfer in variable porosity media. J. Heat Transf. **107**(3), 642–647 (1985)
205. K. Vafai, Analysis of the channeling effect in variable porosity media. J. Energy Res. Technol. **108**(2), 131–139 (1986)
206. A. Amiri, K. Vafai, Analysis of dispersion effects and non-thermal equilibrium, non-Darcian, variable porosity incompressible flow through porous media. Int. J. Heat Mass Transf. **37**(6), 939–954 (1994)
207. A. Amiri, K. Vafai, Transient analysis of incompressible flow through a packed bed. Int. J. Heat Mass Transf. **41**(24), 4259–4279 (1998)
208. W.J. Minkowycz, A. Haji-Sheikh, K.F. Vafai, On departure from local thermal equilibrium in porous media due to a rapidly changing heat source: the Sparrow number. Int. J. Heat Mass Transf. **42**(18), 3373–3385 (1999)
209. B. Alazmi, K. Vafai, Analysis of variants within the porous media transport models. J. Heat Transf. **122**(2), 303–326 (2000)
210. M. Modaresifar, G.J. Kowalski, Numerical simulation of an injection microscale calorimeter to identify significant thermal processes and verify data reduction procedures, in *ASME 2017 International Mechanical Engineering Congress and Exposition* (American Society of Mechanical Engineers, 2017), pp. V008T10A043–V008T10A043
211. A.J. Conway, W.M. Saadi, F.L. Sinatra, G.J. Kowalski, D. Larson, J. Fiering, Dispersion of a nanoliter bolus in microfluidic co-flow. J. Micromech. Microeng. **24**(3), 034006 (2014)
212. M. Modaresifar, G.J. Kowalski, Microscale calorimetric device for determining reaction parameters. Thermochimica Acta **655**, 137–144 (2017)
213. M. Modaresifar, *Thermal Analysis of Chemical Reactions in Microcalorimeter Using Extraordinary Optical Transmission Through Nanohole Arrays*. Ph.D. thesis, Northeastern University (2019)
214. T.N.F. Roach, P. Salamon, J. Nulton, B. Andresen, B. Felts, A. Haas, S. Calhoun, N. Robinett, F. Rohwer, Application of finite-time and control thermodynamics to biological processes at multiple scales. J. Non-Equilib. Thermodyn. **43**(3), 193–210 (2018)
215. H.H. Pennes, Analysis of tissue and arterial blood temperatures in the resting human forearm. J. Appl. Physiol. **1**(2), 93–122 (1948)
216. M.M. Chen, K.R. Holmes, Microvascular contributions in tissue heat transfer. Ann. N.Y. Acad. Sci. **335**(1), 137–150 (1980)
217. S. Weinbaum, L.M. Jiji, D.E. Lemons, Theory and experiment for the effect of vascular microstructure on surface tissue heat transfer-Part I: anatomical foundation and model conceptualization. J. Biomech. Eng. **106**(4), 321–330 (1984)

218. L.M. Jiji, *Heat Conduction: Third Edition*, 3rd edn. (Springer, Berlin, 2009)
219. S. Weinbaum, L.M. Jiji, A new simplified bioheat equation for the effect of blood flow on local average tissue temperature. J. Biomech. Eng. **107**(2), 131–139 (1985)
220. W. Wulff, The energy conservation equation for living tissue. IEEE Trans. Biomed. Eng. **6**(BME-21), 494–495 (1974)
221. H.G. Klinger, Heat transfer in perfused biological tissue-I: general theory. Bull. Math. Biol. **36**, 403–415 (1974)
222. H.G. Klinger, Heat transfer in perfused biological tissue-II: The "macroscopic" temperature distribution in tissue. Bull. Math. Biol. **40**(2), 183–199 (1978)
223. G.T. Anderson, J.W. Valvano, A small artery heat transfer model for self-heated thermistor measurements of perfusion in the kidney cortex. J. Biomech. Eng. **116**(1), 71–78 (1994)
224. A. Zolfaghari, M. Maerefat, A new simplified thermoregulatory bioheat model for evaluating thermal response of the human body to transient environments. Build. Environ. **45**(10), 2068–2076 (2010)
225. A.P. Gagge, Rational temperature indices of man's thermal environment and their use with a 2-node model of his temperature regulation, in *Federation Proceedings*, vol. 32 (1973), pp. 1572
226. P.A. Patel, J.W. Valvano, J.A. Pearce, S.A. Prahl, C.R. Denham, A self-heated thermistor technique to measure effective thermal properties from the tissue surface. J. Biomech. Eng. **109**(4), 330–335 (1987)
227. R.B. Roemer, E.G. Moros, K. Hynynen, A comparison of bioheat transfer and effective conductivity equation predictions to experimental hyperthermia data, *Advances in Bioengineering* (ASME Winter Annual Meeting, 1989), pp. 11–15
228. J.L.M. Hensen, Literature review on thermal comfort in transient conditions. Build. Environ. **25**(4), 309–316 (1990)
229. F. Kreith, *The CRC Handbook of Thermal Engineering* (Springer Science & Business Media, Berlin, 2000)
230. L. Zhu, C. Diao, Theoretical simulation of temperature distribution in the brain during mild hypothermia treatment for brain injury. Med. Biolo. Eng. Comput. **39**(6), 681–687 (2001)
231. G. Chenhua, C. Ruixian, An analytical solution of non-Fourier Chen-Holmes bioheat transfer equation. Chin. Sci. Bull. **50**(23), 2791–2792 (2005)
232. Y-G. Lv, J. Liu, Effect of transient temperature on thermoreceptor response and thermal sensation. Build. Environ. **42**(2), 656–664 (2007)
233. W.J. Minkowycz, E.M. Sparrow, *Advances in Numerical Heat Transfer*, vol. 3. (CRC, Bocca Raton, 2009)
234. A. Zolfaghari, M. Maerefat, Bioheat Transfer. InTech (2011)
235. S. Becker, A. Kuznetsov, *Heat Transfer and Fluid Flow in Biological Processes* (Academic, 2014)
236. C. Truesdell, W. Noll, *The Non-linear Field Theories of Mechanics* (Springer, Berlin, 2004)
237. W. Muschik, Objectivity and frame indifference, revisited. Arch. Mech. **50**(3), 541–547 (1998)
238. T. Matolcsi, P. Ván, Can material time derivative be objective? Phys. Lett. A **353**(2), 109–112 (2006)
239. A. Banerjee, A.A. Ogale, C. Das, K. Mitra, C. Subramanian, Temperature distribution in different materials due to short pulse laser irradiation. Heat Transf. Eng. **26**(8), 41–49 (2005)
240. M. Jaunich, S. Raje, K. Kim, K. Mitra, Z. Guo, Bio-heat transfer analysis during short pulse laser irradiation of tissues. Int. J. Heat Mass Transf. **51**(23), 5511–5521 (2008)
241. P. Dhar, A. Paul, A. Narasimhan, S.K. Das, Analytical prediction of sub surface thermal history in translucent tissue phantoms during plasmonic photo thermotherapy (2015). arXiv:1511.04549
242. D.Y. Tzou, A unified field approach for heat conduction from macro- to micro-scales. J. Heat Transf. **117**(1), 8–16 (1995)
243. D.Y. Tzou, *Macro- to Micro-scale Heat Transfer: The Lagging Behavior* (CRC Press, Bocca Raton, 1996)

244. J. Zhou, J.K. Chen, Y. Zhang, Dual-phase lag effects on thermal damage to biological tissues caused by laser irradiation. Comput. Biol. Med. **39**(3), 286–293 (2009)
245. P. Yuan, Numerical analysis of an equivalent heat transfer coefficient in a porous model for simulating a biological tissue in a hyperthermia therapy. Int. J. Heat Mass Transf. **52**(7), 1734–1740 (2009)
246. P. Hooshmand, A. Moradi, B. Khezry, Bioheat transfer analysis of biological tissues induced by laser irradiation. Int. J. Therm. Sci. **90**, 214–223 (2015)
247. S.A. Rukolaine, Unphysical effects of the dual-phase-lag model of heat conduction. Int. J. Heat Mass Transf. **78**, 58–63 (2014)
248. S.A. Rukolaine, Unphysical effects of the dual-phase-lag model of heat conduction: higher-order approximations. Int. J. Therm. Sci. **113**, 83–88 (2017)
249. M. Fabrizio, B. Lazzari, V. Tibullo, Stability and thermodynamic restrictions for a dual-phase-lag thermal model. J. Non-Equilib. Thermodyn. (2017)
250. M. Fabrizio, B. Lazzari, Stability and second law of thermodynamics in dual-phase-lag heat conduction. Int. J. Heat Mass Transf. **74**, 484–489 (2014)
251. M. Fabrizio, F. Franchi, Delayed thermal models: stability and thermodynamics. J. Therm. Stress. **37**(2), 160–173 (2014)
252. R. Quintanilla, R. Racke, Qualitative aspects in dual-phase-lag heat conduction. Proc. R. Soc. Lond. A: Math. Phys. Eng. Sci. **463**(2079), 659–674 (2007)
253. M. Dreher, R. Quintanilla, R. Racke, Ill-posed problems in thermomechanics. Appl. Math. Lett. **22**(9), 1374–1379 (2009)
254. F. Xu, K.A. Seffen, T.J. Lu, Non-Fourier analysis of skin biothermomechanics. Int. J. Heat Mass Transf. **51**(9), 2237–2259 (2008)
255. N. Sahoo, S. Ghosh, A. Narasimhan, Sa. K. Das, Investigation of non-Fourier effects in bio-tissues during laser assisted photothermal therapy. Int. J. Therm. Sci. **76**, 208–220 (2014)
256. Y. Zhang, Generalized dual-phase lag bioheat equations based on nonequilibrium heat transfer in living biological tissues. Int. J. Heat Mass Transf. **52**(21), 4829–4834 (2009)
257. N. Afrin, J. Zhou, Y. Zhang, D.Y. Tzou, J.K. Chen, Numerical simulation of thermal damage to living biological tissues induced by laser irradiation based on a generalized dual phase lag model. Numer. Heat Transf. Part A: Appl. **61**(7), 483–501 (2012)
258. Kuo-Chi Liu, Han-Taw Chen, Investigation for the dual phase lag behavior of bio-heat transfer. Int. J. Therm. Sci. **49**(7), 1138–1146 (2010)
259. W. Andrä, C.G. d'Ambly, R. Hergt, I. Hilger, W.A. Kaiser, Temperature distribution as function of time around a small spherical heat source of local magnetic hyperthermia. J. Mag. Mag. Mater. **194**(1), 197–203 (1999)
260. D. Tang, N. Araki, N. Yamagishi, Transient temperature responses in biological materials under pulsed IR irradiation. Heat Mass Transf. **43**(6), 579–585 (2007)
261. Y. Miyamoto, W.A. Kaysser, B.H. Rabin, A. Kawasaki, R.G. Ford, *Functionally Graded Materials: Design, Processing and Applications*, vol. 5 (Springer Science & Business Media, Berlin, 2013)
262. B. Kieback, A. Neubrand, H. Riedel, Processing techniques for functionally graded materials. Mater. Sci. Eng.: A **362**(1–2), 81–106 (2003)
263. V. Birman, L.W. Byrd, Modeling and analysis of functionally graded materials and structures. Appl. Mech. Rev. **60**(5), 195–216 (2007)
264. X.-Q. Fang, C. Hu, Dynamic effective thermal properties of functionally graded fibrous composites using non-fourier heat conduction. Comput. Mater. Sci. **42**(2), 194–202 (2008)
265. A. Figueroa, F. Vázquez, Optimal performance and entropy generation transition from micro to nanoscaled thermoelectric layers. Int. J. Heat Mass Transf. **71**, 724–731 (2014)
266. F. Vázquez, A. Figueroa, I. Rodriguez-Vargas, Nonlocal and memory effects in nanoscaled thermoelectric layers. J. Appl. Phys. **121**(1), 014311 (2017)
267. J. Rojas, I. Rivera, A. Figueroa, F. Vázquez, Coupled thermoelectric devices: theory and experiment. Entropy **18**(7), 255 (2016)
268. P. Rogolino, V.A. Cimmelli, Thermoelectric efficiency of graded sicge 1-c alloys. J. Appl. Phys. **124**(9), 094301 (2018)

269. P. Rogolino, A. Sellitto, V.A. Cimmelli, Minimal entropy production and efficiency of energy conversion in nonlinear thermoelectric systems with two temperatures. J. Non-Equilib. Thermodyn. **42**(3), 287–303 (2017)

270. P. Rogolino, A. Sellitto, V.A. Cimmelli, Influence of nonlinear effects on the efficiency of a thermoelectric generator. Zeitschrift für angewandte Mathematik und Physik **66**(5), 2829–2842 (2015)

271. B.-Y. Cao, M. Di Domenico, B.-D. Nie, A. Sellitto, Influence of the composition gradient on the propagation of heat pulses in functionally graded nanomaterials. Proc. R. Soc. A **475**(2221), 20180499 (2019)

272. J. Sladek, V. Sladek, C. Zhang, Transient heat conduction analysis in functionally graded materials by the meshless local boundary integral equation method. Comput. Mater. Sci. **28**(3–4), 494–504 (2003)

273. H.J. Xu, Z.B. Xing, F.Q. Wang, Z.M. Cheng, Review on heat conduction, heat convection, thermal radiation and phase change heat transfer of nanofluids in porous media: Fundamentals and applications. Chem. Eng. Sci. (2018)

274. S. Lee, S.U.-S. Choi, S. Li, J.A. Eastman, Measuring thermal conductivity of fluids containing oxide nanoparticles. J. Heat Transf. **121**(2), 280–289 (1999)

275. J.A. Eastman, S.U.-S. Choi, S. Li, W. Yu, L.J. Thompson, Anomalously increased effective thermal conductivities of ethylene glycol-based nanofluids containing copper nanoparticles. Appl. Phys. Lett. **78**(6), 718–720 (2001)

276. J. Koo, C. Kleinstreuer, A new thermal conductivity model for nanofluids. J. Nanoparticle Res. **6**(6), 577–588 (2004)

277. X.-Q. Wang, A.S. Mujumdar, Heat transfer characteristics of nanofluids: a review. Int. J. Therm. Sci. **46**(1), 1–19 (2007)

278. J. Buongiorno, D.C. Venerus, N. Prabhat, T. McKrell, J. Townsend, R. Christianson, Y.V. Tolmachev, Pa. Keblinski, L.-W. Hu, J.L. Alvarado et al., A benchmark study on the thermal conductivity of nanofluids. J. Appl. Phys. **106**(9), 094312 (2009)

279. J. Eapen, R. Rusconi, R. Piazza, S. Yip, The classical nature of thermal conduction in nanofluids. J. Heat Transf. **132**(10), 102402 (2010)

280. J.-H. Lee, S.-H. Lee, C. Choi, S. Jang, S. Choi, A review of thermal conductivity data, mechanisms and models for nanofluids. Int. J. Micro-Nano Scale Transp. (2011)

281. S.K. Das, S.U.S. Choi, E. Hrishikesh, Patel. Heat transfer in nanofluids–a review. Heat Transf. Eng. **27**(10), 3–19 (2006)

282. P. Keblinski, S.R. Phillpot, S.U.S. Choi, J.A. Eastman, Mechanisms of heat flow in suspensions of nano-sized particles (nanofluids). Int. J. Heat Mass Transf. **45**(4), 855–863 (2002)

283. S.U.S. Choi, Nanofluids: from vision to reality through research. J. Heat Transf. **131**(3), 033106 (2009)

284. S. Kakaç, A. Pramuanjaroenkij, Review of convective heat transfer enhancement with nanofluids. Int. J. Heat Mass Transf. **52**(13–14), 3187–3196 (2009)

285. L. Wang, J. Fan, Nanofluids research: key issues. Nanoscale Res. Lett. **5**(8), 1241 (2010)

286. P. Vadasz, Heat transfer augmentation in nanofluids via nanofins. Nanoscale Res. Lett. **6**(1), 154 (2011)

287. J. Fan, L. Wang, Review of heat conduction in nanofluids. J. Heat Transf. **133**(4), 040801 (2011)

288. V.I. Terekhov, S.V. Kalinina, V.V. Lemanov, The mechanism of heat transfer in nanofluids: state of the art (review). part 1. synthesis and properties of nanofluids. Thermophys. Aeromech. **17**(1), 1–14 (2010)

289. V.I. Terekhov, S.V. Kalinina, V.V. Lemanov, The mechanism of heat transfer in nanofluids: state of the art (review). part 2. convective heat transfer. Thermophys. Aeromech. **17**(2), 157–171 (2010)

290. P. Vadasz, Heat conduction in nanofluid suspensions. J. Heat Transf. **128**(5), 465–477 (2006)

291. J.J. Vadasz, S. Govender, Thermal wave effects on heat transfer enhancement in nanofluids suspensions. Int. J. Therm. Sci. **49**(2), 235–242 (2010)

292. A.A. Mohamad, Myth about nano-fluid heat transfer enhancement. Int. J. Heat Mass Transf. **86**, 397–403 (2015)

293. T. Fülöp, Objective thermomechanics (2015) arXiv:1510.08038

294. P. Ván, G.G. Barnaföldi, T. Bulik, T. Biró, S. Czellár, M. Cieślar, Cs. Czanik, E. Dávid, E. Debreceni, M. Denys et al., Long term measurements from the Mátra Gravitational and Geophysical Laboratory (2018) arXiv:1811.05198

295. T. Fülöp, P. Ván, A. Csatár, Elasticity, plasticity, rheology and thermal stress-an irreversible thermodynamical theory. Elastic **2**(7) (2013)

296. C. Asszonyi, A. Csatár, T. Fülöp. Elastic, thermal expansion, plastic and rheological processes-theory and experiment (2015) arXiv:1512.05863

297. T. Fülöp, Cs. Asszonyi, P. Ván, Distinguished rheological models in the framework of a thermodynamical internal variable theory. Contin. Mech. Thermodyn. **27**(6), 971–986 (2015)

298. M. Szücs, T. Fülöp, Kluitenberg-Verhás rheology of solids in the GENERIC framework. J. Non-Equilib. Thermodyn. **44**(3), 247–259 (2019), arXiv:1812.07052

299. R. Kovács, On the rarefied gas experiments. Entropy **21**(7), 718–730 (2019)

300. R. Kovács, D. Madjarevic, S. Simic, P. Ván, Theories of rarefied gases (2018), ArXiv:1812.10355

301. J. Meixner, Absorption und Dispersion des Schalles in Gasen mit Chemisch Reagierenden und Anregbaren Komponenten. I. Teil. Annalen der Physik **435**(6–7), 470–487 (1943)

302. S.R. de Groot, P. Mazur, *Non-Equilibrium Thermodynamics* (Dover Publications, 1963)

303. M. Carrassi, A. Morro, A modified navier-stokes equation, and its consequences on sound dispersion. Il Nuovo Cimento B **9**, 321–343 (1972)

304. M. Carrassi, A. Morro, Some remarks about dispersion and absorption of sound in monatomic rarefied gases. Il Nuovo Cimento B **13**, 281–289 (1973)

305. D. Jou, C. Perez-Garcia, L.S. Garcia-Colin, M.L. De Haro, R.F. Rodriguez, Generalized hydrodynamics and extended irreversible thermodynamics. Phys. Rev. A **31**(4), 2502 (1985)

306. T. Ruggeri, M. Sugiyama, *Rational Extended Thermodynamics Beyond the Monatomic Gas* (Springer, Berlin, 2015)

307. T. Arima, S. Taniguchi, T. Ruggeri, M. Sugiyama, Dispersion relation for sound in rarefied polyatomic gases based on extended thermodynamics. Contin. Mech. Thermodyn. **25**(6), 727–737 (2013)

308. C.G. Sluijter, H.F.P. Knaap, J.J.M. Beenakker, Determination of rotational relaxation times of hydrogen isotopes by sound absorption measurements at low temperatures I. Physica **30**(4), 745–762 (1964)

309. E. Meyer, G. Sessler, Schallausbreitung in gasen bei hohen frequenzen und sehr niedrigen drucken. Zeitschrift für Physik **149**, 15–39 (1957)

310. H. Struchtrup, *Macroscopic Transport Equations for Rarefied Gas Flows* (Springer, Berlin, 2005)

311. H. Struchtrup, Resonance in rarefied gases. Contin. Mech. Thermodyn. **24**(4–6), 361–376 (2012)

312. J.E. Rhodes Jr., The velocity of sound in hydrogen when rotational degrees of freedom fail to be excited. Phys. Rev. **70**(11–12), 932 (1946)

313. M. Greenspan, Propagation of sound in five monatomic gases. J. Acoust. Soc. Am. **28**(4), 644–648 (1956)

314. C.G. Sluijter, H.F.P. Knaap, J.J.M. Beenakker, Determination of rotational relaxation times of hydrogen isotopes by sound absorption measurements at low temperatures II. Physica **31**(6), 915–940 (1965)

315. T. Arima, S. Taniguchi, T. Ruggeri, M. Sugiyama, Extended thermodynamics of dense gases. Contin. Mech. Thermodyn. **24**(4–6), 271–292 (2012)

316. T. Arima, S. Taniguchi, T. Ruggeri, M. Sugiyama, Extended thermodynamics of real gases with dynamic pressure: an extension of Meixner's theory. Phys. Lett. A **376**(44), 2799–2803 (2012)

317. T. Arima, T. Ruggeri, M. Sugiyama, S. Taniguchi, Non-linear extended thermodynamics of real gases with 6 fields. Int. J. Non-Linear Mech. **72**, 6–15 (2015)

318. H. Struchtrup, P. Taheri, Macroscopic transport models for rarefied gas flows: a brief review. IMA J. Appl. Math. **76**(5), 672–697 (2011)
319. H. Struchtrup, Stable transport equations for rarefied gases at high orders in the Knudsen number. Phys. Fluids **16**(11), 3921–3934 (2004)
320. H. Struchtrup, M. Torrilhon, Higher-order effects in rarefied channel flows. Phys. Rev. E **78**(4), 046301 (2008)
321. J.A. Gracki, G.P. Flynn, J. Ross, Viscosity of Nitrogen, Helium, Hydrogen, and Argon from -100 to 25 c up to 150-250 atmospheres. *Project SQUID Technical Report*, p. 33 (1969)
322. J.A. Gracki, G.P. Flynn, J. Ross, Viscosity of Nitrogen, Helium, Hydrogen, and Argon from -100 to 25 c up to 150–250 atm. J. Chem. Phys. **9**, 3856–3863 (1969)
323. J.H. Dymond, Corrections to the Enskog theory for viscosity and thermal conductivity. Phys. B **144**(3), 267–276 (1987)
324. W.M. Haynes, Viscosity of gaseous and liquid argon. Physica **67**(3), 440–470 (1973)
325. A. Van Itterbeek, W.H. Keesom, Measurements on the viscosity of helium gas between 293 and 1.6 k. Physica **5**(4), 257–269 (1938)
326. A. Van Itterbeek, A. Claes, Measurements on the viscosity of hydrogen-and deuterium gas between 293 K and 14 K. Physica **5**(10), 938–944 (1938)
327. A. Van Itterbeek, O. Van Paemel, Measurements on the viscosity of gases for low pressures at room temperature and at low temperatures. Physica **7**(3), 273–283 (1940)
328. D. Sette, A. Busala, J.C. Hubbard, Energy transfer by collisions in vapors of chlorinated methanes. J. Chem. Phys. **23**(5), 787–793 (1955)
329. P.S. Van der Gulik, C.A. ten Seldam, Density dependence of the viscosity of some noble gases. Int. J. Thermophys. **23**(1), 15–26 (2002)
330. R. Umla, V. Vesovic, Viscosity of liquids-Enskog-2σ model. Fluid Phase Equilib. **372**, 34–42 (2014)
331. Y. Cohen, S.I. Sandler, The viscosity and thermal conductivity of simple dense gases. Ind. Eng. Chem. Fundam. **19**(2), 186–188 (1980)
332. P.S. Van der Gulik, N.J. Trappeniers, The viscosity of argon at high densities. Phys. A: Stat. Mech. Appl. **135**(1), 1–20 (1986)
333. P.S. Van der Gulik, N.J. Trappeniers, Application of Enskog theory on the viscosity of argon. Phys. B+C **139**, 137–139 (1986)
334. V.K. Michalis, A.N. Kalarakis, E.D. Skouras, V.N. Burganos, Rarefaction effects on gas viscosity in the Knudsen transition regime. Microfluid. Nanofluid. **9**(4–5), 847–853 (2010)
335. A. Beskok, G.E. Karniadakis, Report: a model for flows in channels, pipes, and ducts at micro and nano scales. Microscale Thermophys. Eng. **3**(1), 43–77 (1999)
336. E. Roohi, M. Darbandi, Extending the Navier-Stokes solutions to transition regime in two-dimensional micro-and nanochannel flows using information preservation scheme. Phys. Fluids **21**(8), 082001 (2009)

Chapter 6
Notes on the Solutions of PDE Systems—Duality Between Two Worlds

1 The Beginning of the End

So far, we have intentionally omitted the presentation of the solution methods of partial differential equations. To obtain a physically relevant solution, one has to be aware of the characteristics of the problem, even before trying to solve it. This is important due to the reasons emphasized at the very beginning of the modeling aspects:

- fixing the required variables and choosing the appropriate constitutive equation(s),
- defining the initial and boundary conditions,
- finding the relevant time and length scales,

moreover, all these aspects are extended with the particular properties of the chosen solution method. There are two possibilities to solve a system of PDEs: analytically or numerically.

The analytical methods yield a solution in a closed form, that is, obtained by 'hand on paper'. Unfortunately, that method is severely restricted. In most cases, the problem must be linearized and simplified to be one-dimensional with steady boundary conditions. Naturally, there are exceptions, in some exceptional and rare cases, it is possible to solve non-linear problems with time-dependent boundary conditions; however, it cannot be done generally.

On the contrary, the numerical methods allow to solve much more complicated problems, even the simplest approach can be helpful. However, these algorithms provide only approximate solutions.[1] Furthermore, the discrete (numerical) model suffers from other problems which are not part of the real one, such as numerical dispersion and damping. For instance, it is easy to obtain numerically a damped wave solution for the wave equation that consists of no damping term. This is the characteristic property of the applied algorithm only and must be realized before accepting the solution as a real physical one.

---

[1]In some cases, it is possible to obtain an exact solution, too. For instance, see numerical integration [1].

© Springer Nature Switzerland AG 2020
V. Józsa and R. Kovács, *Solving Problems in Thermal Engineering*, Power Systems,
https://doi.org/10.1007/978-3-030-33475-8_6

This chapter is devoted to present the essential steps of the way to obtain a physically sound solution. Since the methods for classical models such as the Fourier, Fick, and Navier–Stokes equations are well-developed [2–6], we will focus our attention on the non-classical ones. These models require an improvement about how to think about the boundary conditions and how to implement them into the numerical approach. However, the following ideas and techniques are not limited to the generalized models, they can be automatically applied to the conventional problems, too. Thus first, we start with discussing the boundary conditions, and then we can turn our focus on the discretization methods.

2   Notes on Thermal Boundary Conditions

The usual boundary conditions of the Fourier heat equation are well-known in the engineering practice. It is essential to emphasize: all these boundary conditions are formulated using the temperature T as a basic field variable in the Fourier heat equation,

$$\frac{\partial T}{\partial t} = \alpha \Delta T, \tag{1}$$

with α being the thermal diffusivity with constant material parameters when heat generation is absent. The corresponding boundary conditions are:

- Dirichlet-type (first kind) boundary: $T(t)$ is given on the boundary.
- Neumann-type (second kind) boundary: the gradient of the basic field variable is given, i.e., $\nabla T(t)$ is prescribed.
- Robin-type (third kind) boundary: the linear combination of the previous boundary conditions. One commonly used case prescribes heat convection in the normal direction of the surface: $\mathbf{q} \cdot \mathbf{n} = -\lambda \nabla T \cdot \mathbf{n} = h(T - T_\infty)$ with $\mathbf{n}$ being the normal vector of the surface element and h is the heat transfer coefficient.

Naturally, there are other boundary conditions as special cases of the latter such as the thermal radiation ($\mathbf{q} \cdot \mathbf{n} = \beta(T^4 - T_\infty^4)$) and symmetry[2] ($\mathbf{q} \cdot \mathbf{n} = 0$). While the temperature T is our basic field variable, all the previous boundary conditions can be applied conveniently during numerical calculations. More importantly, their definitions are the consequence of the constitutive equation:

$$\mathbf{q} = -\lambda \nabla T, \tag{2}$$

that connects the temperature gradient to the heat flux. In more general situations, that relationship does not hold, and it also unavoidably effects the definition of boundary

---

[2]Classically, the symmetry condition is defined by $\nabla T = 0$ which means symmetric temperature field, it can be easily imagined for a one-dimensional case. However, for non-Fourier heat equations, that definition is not general enough.

conditions. Consequently, it is not possible to use the interpretation of boundaries, as mentioned above. This is crucial in the modeling of non-classical phenomena. In general, it is more natural to define the boundaries for the heat flux instead of the temperature. Moreover, they are rarely constant and mostly depend on time. Then, even dealing with Fourier's law, it is more convenient to reformulate the boundary condition to be the first kind. It is possible by changing the primary variable to the heat flux. In this case, the Fourier heat equation is equivalent with

$$\frac{\partial \mathbf{q}}{\partial t} = \alpha \Delta \mathbf{q}, \tag{3}$$

which form appears to be useful in analytical solutions, too. It is worth to keep in our mind that the complete heat equation reads

$$\rho c \frac{\partial T}{\partial t} + \nabla \cdot \mathbf{q} = 0, \quad \mathbf{q} = -\lambda \nabla T, \tag{4}$$

in which the balance equation is separated from the constitutive equation. In case of the presence of source terms in the balance equation, it becomes more straightforward how to handle both analytically and numerically the problem, especially when one would like to eliminate one of the variables. Nevertheless, the balance equation is necessary: the temperature field must be reconstructed due to practical reasons.

Using $\mathbf{q}$ as a field variable has significant consequences to the boundary conditions: it is no longer possible to prescribe the temperature on the boundary. Instead, the time dependence of the heat flux is more comfortable to handle in analytical solutions. Then the temperature field can be easily reconstructed by integration. Later, we will recall that strategy, discussing a particular numerical approach.

That is, thermodynamics allows to customarily choose the primary field variable, which one fits better to the actual task. Practically, it does not merely allow to handle the boundaries easier, but, in parallel, also permits to 'eliminate the boundaries' for the other field variables. It will be demonstrated soon in various examples. In summary, it is not trivial how many and what kind of boundary conditions are needed to solve a PDE system.

In the case of the Fourier heat equation, the relationship between the heat flux and the temperature is both mathematically and physically clear. Although this is the most often occurring situation, it is not satisfactory in the modeling of non-classical phenomena. For non-Fourier heat conduction, such a simple relationship does not exist anymore, i.e., the constitutive equation becomes a time evolution equation for the heat flux that may contain spatial derivatives as well. For example, the Guyer–Krumhansl equation[3] in temperature representation reads

$$\tau \frac{\partial^2 T}{\partial t^2} + \frac{\partial T}{\partial t} = \alpha \Delta T + l^2 \frac{\partial(\Delta T)}{\partial t}. \tag{5}$$

---

[3] See Chap.2 for details.

Since the corresponding constitutive equation is

$$\tau \frac{\partial \mathbf{q}}{\partial t} + \mathbf{q} = -\lambda \nabla T + l^2 \Delta \mathbf{q}, \tag{6}$$

it is not possible to directly connect the temperature to the heat flux. Consequently, a time-dependent heat flux $\mathbf{q}(t)$ cannot be formulated[4] with the temperature $T(t)$ due to the non-local term on the right hand side; it requires the knowledge of the spatial behavior in advance.[5]

Now, let us think about the number of necessary boundary conditions. Usually, the highest order of the spatial derivative gives that number which is trivial for a single differential equation. However, for a system of partial differential equations, it is not straightforward. When one uses two field variables (e.g., the temperature and the heat flux), then, for the Fourier model (see Eq. (4)), the above mentioned 'rule of thumb' suggests 1 condition for the temperature and 1 for the heat flux. For the Guyer–Krumhansl equation, that way of thinking suggests 1 condition for the temperature and 2 for the heat flux. It is not conclusive for PDE systems and gives no real clue about the appropriate boundaries. Indeed, that question can be answered if one considers only the primary field variable, e.g., the heat flux. That is, 2 conditions for the heat flux is enough in 1D, both for the Fourier and Guyer–Krumhansl models. As our experience in numerical modeling shows, this is suitable to obtain a real, physically acceptable solution. This is validated using the previously mentioned analytical approach.

Overall, in this chapter, we show an example to the analytical solution of Guyer–Krumhansl equation, in which the boundary conditions are defined according to the heat pulse experiment. It demonstrates the importance of these aspects [7]. Although the mentioned analytical solution is only applicable in a significantly simplified case, it can be still useful to validate the numerical solutions. Regarding the numerical methods, solely the most important aspects are discussed, in close connection to the boundary conditions.

As a closing remark, we emphasize the universality of these general principles. They hold for coupled models as well such as the Navier–Stokes–Fourier equations. In such a situation, more than one primary field variable should be used. That choice is made arbitrarily, suitably fitted to the actual task.

2.1   Notes on Initial Thermal Conditions

One must represent the initial state of the body, including not merely the current value of the state variables but maybe their time derivatives as well. In the case of a classical heat conduction problem, it means that the temperature field is given at the

---

[4]It is still an open mathematical question.

[5]As an analogy: think about the Brinkman equation, the non-local generalization of Darcy's law.

initial time instant and that temperature field restricts the heat flux up to an additive constant. For non-Fourier heat conduction, that question is a bit more complicated as there are multiple possibilities.

First, when solving the system of PDEs numerically, it is more convenient to use all the variables as the boundary conditions can be more suitably defined in this way. Usually, a PDE system consists of first-order equations in time, thus it is enough to prescribe the initial value of all variables. For example, solving the Guyer–Krumhansl heat equation by treating the balance of internal energy and the constitutive equation separately, the initial values of the state variables (T and $\mathbf{q}$) are prescribed, that is, 2 initial conditions are given.

However, when one intends to obtain an analytical solution, it could be easier to use only one variable instead of the complete system. Hence, a single PDE must be solved, which one contains higher-order time derivatives, too.[6] That approach dictates that besides the initial value of the chosen variable, one must characterize its time derivatives as well. Since this corresponding variable is chosen according to the available boundary conditions, the resulting PDE is solvable. Then, reconstructing the other fields, further initial conditions may be necessary. For a demonstration, let us consider the heat flux representation of the Guyer–Krumhansl equation:

$$\tau\frac{\partial^2\mathbf{q}}{\partial t^2} + \frac{\partial\mathbf{q}}{\partial t} = \alpha\Delta\mathbf{q} + l^2\frac{\partial(\Delta\mathbf{q})}{\partial t}. \tag{7}$$

Its solution requires 2 initial conditions for the heat flux:

1. given initial value, $\mathbf{q}(\mathbf{x}, t = 0) = f(\mathbf{x})$,
2. given time derivative, $\partial_t\mathbf{q}(\mathbf{x}, t = 0) = g(\mathbf{x})$,

in which the functions f and g are not independent of each other since both of them must satisfy the constitutive equation. Consequently, it is not possible to arbitrarily choose them. It becomes more apparent by using the complete set of variables instead of a single one. Having $\mathbf{q}$, the temperature field can be calculated by integrating the heat flux, and that integration requires an initial value for the temperature, too. Overall, such an approach requires 3 initial conditions at the end, instead of 2. However, for non-zero f, the initial temperature field may be restricted, and that conditions are not independent of each other. The safest option is always to choose homogeneous fields for initial conditions.

3 Notes on the Analytical Solutions

The analytical solutions are not necessarily exact but could provide some formula in a closed-form. Even the approximate solutions are valuable as they can offer an insight into the behavior of the PDE system, helping the adequate choice of a

---

[6]For instance, see the temperature representation of the Guyer–Krumhansl equation (5).

numerical method. Usually, only one-dimensional problems are investigated analytically, in higher dimensions the task becomes almost impossible, especially with time-dependent boundary conditions.[7] Since the analytical approach of non-Fourier equations requires a modern interpretation of boundary conditions, their literature is not as elaborated as of the Fourier model. Here, we summarize the most important aspects and show the indispensable steps of solving the Guyer–Krumhansl equation with heat pulse boundary conditions, according to [7]. The classical mathematical techniques can be found in [4, 9, 10].

First, let us formulate the problem statement,

$$\rho c \partial_t T + \partial_x q = 0, \qquad\qquad \tau \partial_t q + q + \lambda \partial_x T - l^2 \partial_{xx} q = 0, \qquad (8)$$

$$T(x, t = 0) = T_0, \qquad\qquad\qquad\qquad q(x, t = 0) = 0, \qquad (9)$$

in which Eq. (8) is the one-dimensional form of the Guyer–Krumhansl heat equation, and Eq. (9) presents the initial conditions. The heat pulse boundary is modeled with

$$q(x = 0, t) = q_0(t) = \begin{cases} q_{max} \left(1 - \cos \left(2\pi \cdot \frac{t}{t_p} \right) \right) & \text{if } 0 < t \le t_p, \\ 0 & \text{if } t > t_p, \end{cases} \qquad (10)$$

that function is convenient from numerical point of view since its derivatives respect to t are zero at $t = 0$ and $t = t_p$, where t_p is the pulse length. The rear side boundary is considered to be adiabatic[8] ($q(x = L, t) = q_L(t) = 0$). Now the Eqs. (8)–(10) form the problem statement together. Its well-posedness, that requires existence, uniqueness and continuous dependence on initial conditions, is not proved here.

It is more suitable to use the heat flux representation of the Guyer–Krumhansl equation since we can prescribe the heat flux according to the experimental setup. Thus we use the Eq. (7), simplified to one spatial dimension. Furthermore, the front side boundary condition changes after t_p, therefore it is convenient to split the whole solution into two sections: $q_I(x, t \le t_p)$ and $q_{II}(x, t > t_p)$. We begin with the first one.

Since one boundary condition is not identically zero, it stands as an 'inhomogeneity' and must be homogenized[9] as follows. The decomposition

$$q_I(x, t) = v(x, t) + w(x, t) \qquad (11)$$

allows to separate the 'inhomogeneity' from the homogeneous solution $v(x, t)$. The simplest choice for w is

$$w(x, t) := q_0(t) + \frac{x}{L} \big(q_L(t) - q_0(t) \big). \qquad (12)$$

---

[7] Naturally, there are exceptions, for instance, see the book of Ozisik [8].

[8] In more realistic situations, the convective heat transfer also should be considered.

[9] Also, the heat generation terms, if present.

Then, for instance, it is possible to calculate v using the method of separation of variables, i.e.,

$$v(x, t) = \varphi(t)X(x), \tag{13}$$

where $X(x)$ are the eigenfunctions of the Laplacian, obtained as a solution of the Sturm–Liouville problem, accounting the boundary conditions as well. The complete set of $X(x)$ forms an orthonormal basis and the solutions of q are represented within. Regarding $\varphi(t)$, one has to solve a second order ODE, using the initial conditions for q, wherein the time derivative of q is also zero and the decomposition of q does not effectuate it, $q_I(x, t = 0) = v(x, t = 0) + w(x, t = 0) = 0$. Having the solution for q_I, it is possible to reconstruct the corresponding temperature distribution T_I, using the balance of internal energy,

$$T_I(x, t) = -\frac{1}{\rho c} \int_{t_0=0}^{t} \partial_x q_I(x, \tilde{t}) d\tilde{t}, \ (t \leq t_p). \tag{14}$$

Repeating the previous calculation for the section II is straightforward, however, we draw the attention to the intial conditions, since q_{II} and T_{II} continue the first one and may not be trivial to account them appropriately.

In order to be correct, we must make some remarks here.

1. The $q_0(t)$ boundary condition is advantageous both from the numerical and analytical point of view due to its smoothness. Not any $q_0(t)$ can be applied here, for instance, using the Heaviside function breaks the differentiability around the jump.
2. In fact, the solution of $\varphi(t)$ may not be as easy as it seems to be. One consequence of the Eq. (11) appears here as an inhomogeneity. We assumed that this term could be represented within the same basis, constructed from $X(x)$.
3. Although other techniques exist (Laplace and Fourier transformation, asymptotic waveform evaluation, and so on), the separation of variables is still applicable and does not suffer from difficulties occurring during the inverse transformation.
4. The role of the boundary conditions is essential. Even the smallest incompatibility can lead to physically contradictory results such as the violation of maximum principle [11–14].

4 Notes on the Numerical Methods

The discretization is a process that allows representing continuous quantities on a discrete lattice. The distance between each grid point is called 'space step'. Consequently, only the spatial operators[10] can be approximated on a lattice and this is

---

[10]Here we restrict ourselves on derivatives.

called spatial discretization. A time step updates the values on the lattice. It is important to emphasize that physical quantities are not necessarily placed only on the grid points but can be found between them as well. The approximations of time and space derivatives can be made independently, but both of them influence the outcome of the numerical solution.

In the literature, one can find numerous numerical methods, each can be characterized by its precision, stability properties, possible numerical errors, and others. The overall applicability of a scheme depends on these properties and, eventually, one must choose a method accordingly, which fits the best to the actual physical problem.

Thus, even before choosing a numerical method, one should pay attention to the possible physical content of the solution. For instance, the modeling of an elastic wave propagation requires a scheme that preserves the energy in every time step and free from artificial dissipation. Since every field of physics has its peculiarity, it seems to be impossible to develop a universal numerical method that can be applied for every physical problem. Overall, one must make some compromise among the following aspects:

- quickness,
- stability and robustness,
- low dissipation and dispersion errors,
- high precision, fast convergence,
- easy realization of the algorithm, possibility to be parallelized,
- reflects the appropriate physical content,
- easy implementation for practical problems.

No scheme can fulfill all these aspects simultaneously. For example, the finite element methods (FEM) easily handle almost any type of geometry, quickly solves stationary problems even for dissipative systems, but not efficient for hyperbolic tasks such as wave propagation.[11] On the contrary, the finite difference methods are easy to construct and realize, also applicable for hyperbolic equations, but their precision is limited.

A reliable numerical solution requires to prove its convergence and validate by an experiment and/or an analytical solution. Unfortunately, none of the validation processes are trivial, it demands a lot of effort and stands as an unavoidable step. Reproducing the same solution using another numerical method does not ensure the convergence. In summary, the most critical requirement is the physical content of the solution. Later, we show an example in which it is easy to obtain a stable solution that has zero physical relevance. Since these properties are not apparent, especially when the problem is complex, one has to pay increased attention.

The outcome of the discretization process is a system of algebraic equations. Its solution is not intended to present here. We refer to the well-known literature about the details [1, 15, 16]. This section aims to present the peculiarities of the relevant methods, emphasizing the implementation of boundary conditions.

---

[11] In fact, not just simply not 'efficient' but mostly incapable.

4.1 Finite Elements in Thermal Problems

Practically, the whole geometry is divided into 'elements', and the spatial discretization is the 'meshing' process. Mathematically, it transforms the PDE into a coupled system of ordinary differential equations (ODE). Then the resulting ODEs must be integrated in time; however, this is not part of this method.

There are numerous books in the literature [6, 17–19] in which the details of finite element analysis is deeply discussed. The emphasis is mostly on the mechanical part. Instead of repeating their presentation, we choose to briefly summarize the relevant terms in the resulting finite element equations. A transparent derivation is presented in the ANSYS theory guide, which could be useful to the advanced users of finite element software. The derivation of thermal finite element equations aims to obtain a similar system in the analogy of mechanics. In the following, we present the essential steps and notions only.

All elements are characterized by their shape function N, which interpolates between the nodes, and mainly constraints the behavior of an element. It also plays a crucial role in the modeling of distributed loads such as volumetric heat generation. In the following, let us assume only one element with two nodes. Using this formalism, the temperature is represented as a vector that contains the nodal values of the element. Moreover, in order to keep the notations clear as much as possible, the index notation is used with Einstein summation convention. That is, as the temperature is a vector, and hence denoted as T_e^i.[12] The subscript e indicates the 'elemental' quantities, a vector or a matrix that characterizes a single element. The corresponding vector of shape functions is N^i, and thus, the nodal temperature is $T = N^i T_e^i$. The temperature gradient is $\nabla T = \partial^i T = \partial^i N^j T_e^j = B^{ij} T_e^j$ where B^{ij} is the gradient of the shape functions. Since the shape functions depend only on spatial coordinates, the time derivative of the temperature is $\partial_t T = N^i \partial_t T_e^i$. Then, we need the definition of 'virtual temperature', $\delta T = \delta T_e^i N^i$, which is the set of permissible temperature distributions. This is an analog notion with the virtual displacement.[13]

For the sake of completeness, we consider the Fourier heat equation for a moving media, in which the substantial time derivative appears on the left-hand side with the velocity v^i,

$$\rho c(\partial_t T + v^i \partial^i T) = \partial^i(\lambda^{ij}\partial^j T) + q_E, \tag{15}$$

where q_E stands for the overall inlet and outlet heat flux on the element, including the volumetric heat generation and the convective heat transfer effects, too. Moreover, in the most general case, λ^{ij} is a matrix of thermal conductivities that is characteristic for an anisotropic media. Now, multiplying the Eq. (15) with the virtual temperature

---

[12]In general, there is a distinguished meaning of lower and upper indices both in physics and mathematics. Here we use only upper indices and we do not distinguish vectors from covectors.

[13]In the field of meteorology, the virtual temperature has an entirely different meaning: this is the temperature of the dry air in case if it has the same pressure and density as of the moist air [20].

δT, and integrating respect to the overall volume of the corresponding element, it yields

$$
\int_V \rho c \delta T_e^i N^i N^j \partial_t T_e^j dV + \int_V \rho c \delta T_e^i N^i v^j B^{jk} T_e^k dV + \int_V \delta T_e^i B^{ij} \lambda^{jk} B^{kl} T_e^l dV =
$$

$$
= \int_S \delta T_e^i N^i q_A dS + \int_S \delta T_e^i N^i h(T_\infty - N^j T_e^j) dS + \int_V \delta T_e^i N^i q_V dV,
$$

$$
\tag{16}
$$

in which term of q_E is splitted into a surface, a convective and a volumetric contributions. Since the surface heat flux q_A is not necessarily proportional with the temperature, the convective part is treated separately. Then let us consider the following simplification assumptions:

- the mass density ρ and the specific heat c are constants,
- the heat transfer coefficient h and the environment temperature T_∞ are also constants,
- as the quantities T_e^i, $\partial_t T_e^i$ and δT_e^i are nodal values, they can be taken out and both sides can be simplified with δT_e^i. That is, it is not a real variational principle, on contrary to the usual approach of finite element methods to mechanical problems.

Finally, the equation reads

$$
\rho c \int_V N^i N^j dV \partial_t T_e^j + \rho c \int_V N^i v^k B^{kj} dV T_e^j + \int_V B^{ik} \lambda^{kl} B^{lj} dV T_e^j =
$$

$$
= \int_S N^i q_A dS + \int_S h T_\infty N^i dS - \int_S h N^i N^j dS T_e^j + \int_V q_V N^i dV, \tag{17}
$$

where we can introduce the following quantities:

- $C_c^{ij} := \rho c \int_V N^i N^j dV$, this is the heat capacity matrix of one element, it is also called thermal damping matrix highlighting the mechanical analogy behind the finite element formalism. It is mostly a diagonal matrix but not for every element type. The diagonalization procedure introduces some errors that lead to decreasing heat capacity (analogously, mass decreasing in mechanics).
- $K_m^{ij} := \rho c \int_V N^i v^k B^{kj} dV$, which is a non-symmetric matrix that is related to the convective mass transport.
- $K_d^{ij} := \int_V B^{ik} \lambda^{kl} B^{lj} dV$, called diffusion matrix. However, its name is misleading as it consists the thermal conductivities instead of thermal diffusivities.
- $K_c^{ij} := \int_S h N^i N^j dS$, called convection matrix which is describes the convective heat transport on the boundary.

- $Q^i := \int_S N^i q_A dS$, according to terminology, it is often called as "mass flux" vector that simply follows the mechanical analogy. Used to define the heat flux boundary condition.
- $Q_c^i := \int_S hT_\infty N^i dS$, called surface convection heat flow vector and used to define the convection boundary condition.
- $Q_g^i := \int_V q_V N^i dV$, which is the vector of internal heat generation.

Using these quantities, Eq. (17) reduces to

$$C_c^{ij} \partial_t T_e^j + (K_m^{ij} + K_d^{ij} + K_c^{ij}) T_e^j = Q^i + Q_c^i + Q_g^i, \tag{18}$$

which one is practically a system of coupled ODEs, wherein one ODE describes the time evolution of a nodal temperature. Therefore, in the simplest case[14] two coupled ODEs must be solved. The coupling is ensured through the K matrices, which are to characterize the influence of one node to the others. Eq. (18) can be dramatically simplified, assuming $v^i = 0^i$, with no convective heat transport. In general, linear shape functions are used to formulate these matrices. However, they are mostly inappropriate in situations with volumetric heat generation due to the nonlinear nature of the temperature distribution.

The thermal loading (heat flux and convection) boundary conditions are prescribed in the particular elements of the vectors Q, on the right-hand side. The first-type boundaries mean fixed values in the nodal temperature vector. The finite element formulations of generalized heat equations (e.g., the Maxwell–Cattaneo–Vernotte and the dual phase lag equations) is not easy but straightforward [21–27]. The difficulties start with the implementation of non-local terms, e.g., with the Laplacian of the heat flux such as the Guyer–Krumhansl equation does describe. As it is mentioned earlier, it is not possible to prescribe both the temperature and the heat flux on the boundary for the Guyer–Krumhansl equation. However, on the contrary, all field variables must be presented in the nodal degrees of freedom in the finite element approach, which makes extraordinarily challenging to obtain real, physically admissible solutions [28].

4.2 Finite Differences for PDE Systems

The approximation of derivatives with finite differences is a well-known numerical method and is among the simplest ones. On one hand, despite its simplicity, it applies to a wide range of engineering problems. On the other hand, only simple geometries can be discretized with equidistant lattice.[15] This section intends to present

---

[14]One-dimensional problem, solved with one element, using linear shape functions.

[15]That is, the spatial steps cannot be varied within one direction. In contrary, the time steps can be programmed to be adaptive.

some essential techniques to solve PDE systems. We show examples of thermal and mechanical models as well, in which the time-stepping strategy is emphasized, besides the implementation of boundary conditions.

The difference schemes exploit the Taylor series expansion and result in order of approximation according to the order of truncation. Thus, it is easy to estimate the precision and order of convergence. Moreover, the consistency[16] is also easily fulfilled [28, 29]. The fulfillment of consistency and stability[17] together result in convergence, according to the Lax theorem [30, 31]. In the following, we present the simplest (and most often used) difference schemes. Their detailed discussion and derivation can be found in [1, 16]. When discretizing a time derivative, the two basic options are the following:

- forward discretization: $\frac{\partial f}{\partial t} \approx \frac{f_j^{n+1} - f_j^n}{\Delta t}$,
- backward discretization: $\frac{\partial f}{\partial t} \approx \frac{f_j^n - f_j^{n-1}}{\Delta t}$,

with $f = f(x, t)$ being an arbitrary, differentiable function, n is the index of time steps and j denotes the spatial steps. Both of these approximations are first order and related to the jth point. Naturally, one could choose a different approximation with significantly higher precision such as the Adams–Bashforth and Runge–Kutta schemes [32, 33], however, for our goals, it is not necessary. As we will see later, even the simplest methods can work.[18]

Regarding the spatial discretization, the previous approximations are valid options to choose, realizing the differentiation respecting to the index j. The third one is called central discretization:

$$\frac{\partial f}{\partial x} \approx \frac{f_{j+1}^n - f_{j-1}^n}{2\Delta x}, \tag{19}$$

because the neighbouring points of j are used, and it results in a second order approximation.

Since the Taylor series expansion goes to infinity, it is possible to derive infinitely many approximations, resulting in higher and higher precision. However, their applicability is strongly limited from a practical point of view. Regarding the time derivatives, one must find initial conditions as many as time instants are used in the scheme. This is the smaller problem as those states can be safely ignored with initial equilibrium conditions.[19] The situation is different in the spatial direction: the imple-

---

[16]We distinguish weak and strong consistency. For the weak consistency, the requirements are simple: when $\Delta x \to 0$ and $\Delta t \to 0$, then $D_x \to \partial_x$ and $D_t \to \partial_t$, where D_x and D_t are the discrete (difference) operators. In other words, the discrete system converges to the continuous one. In general, this is not satisfactory to the strong consistency, which means that all the properties of the PDE system are inherited to the discrete one.

[17]We discuss the corresponding properties soon.

[18]Actually, starting with the simplest approximation is the safest option in order to understand the nature of the PDE system. Besides, the most precise schemes are not the best choice sometimes.

[19]In any other case, choosing a small time step size, one can obtain the initially needed information with a lower order scheme.

mentation of boundary conditions acts as a stronger constraint. Near the boundary, there is not enough grid point to calculate the chosen approximation, solely using 'virtual' points. However, these 'virtual' points can represent only artificially created states since they are over the boundary, i.e., beyond the real domain of interest. Their application is reasonable only for symmetry conditions.[20] Moreover, applying a different scheme, even in the vicinity of the boundary, makes the complete calculation unstable. In summary, high order approximations for spatial derivatives are rarely used. Furthermore, the overall precision of the numerical calculation is given by the lowest order approximation. Consequently, using a second-order scheme for the spatial derivative makes it unreasonable to apply a fourth-order time-stepping algorithm. We want to emphasize the following remarks:

- The possible combinations of the approximations of time and spatial derivatives can be divided into two groups. In the first one, to calculate the new $(n + 1)$th value, only values from the previous step are used. These are called 'explicit' schemes. Their counterpart is called 'implicit' schemes in which the calculation also requires values from the new time step. It essentially determines the stability properties.

- Not any combination of the time and spatial approximations are efficient for all PDE. For instance, the so-called FTCS (forward time centered space) scheme cannot solve the advection equation:

$$\frac{\partial f}{\partial t} = v \frac{\partial f}{\partial x}. \tag{20}$$

Instead, for example, applying Lax's method (i.e., changing f_j^n to $1/2(f_{j+1}^n + f_{j-1}^n)$) is more suitable to obtain a stable (and convergent) solution with the famous CFL (Courant–Friedrich–Lewy) stability criterion: $\frac{v\Delta t}{\Delta x} < 1$. This stability criterion is not a general one, it may work merely for hyperbolic PDEs. In the next section, we show a method to calculate the stability conditions.

- The instability of a scheme for a particular PDE does not mean that this is the case for any PDE using the same scheme. It means that a particular approximation is not suitable for a particular PDE.

- The CFL condition highlights the role of the time scales: the ratio of the chosen time and space steps represent a sort of 'discrete velocity' ($\Delta x/\Delta t$), and compares to the physical one (v). That is, the discretization must respect the present propagation speeds and must be adapted to them to obtain an acceptable resolution of the modeled phenomenon. In other words, the discrete velocity must keep up with the physical one. In some cases, v is around the speed of sound, or even faster.[21] Regarding the ODEs, this behavior is called 'rigidity' and the equation itself is called 'rigid'. The appearance of multiple time scales in the same model makes more difficult the calculations as always the lowest one restricts the maximum

---

[20] We note here again: in thermal problems, considering $\nabla T = 0$ as a symmetry condition is valid only for the Fourier's law.

[21] In electrodynamics, usually, the speed of light dictates the relevant time scales.

applicable time step. There are two ways to overcome this difficulty. The first option is to choose a numerical method, usually an implicit one, that allows to use higher time steps and practically ignoring the processes which occur on that time scale. It is only possible when it offers an acceptable precision. The other option is to rescale the equations with respect to the fastest phenomenon.

Concept of Staggered Fields

Here, we show an efficient way for spatial discretization, which respects the boundary conditions and also eases their implementation. This is extremely important for generalized equations that contain nonlocal terms as well, such as the Guyer–Krumhansl equation, but not restricted to thermal problems in principle. Although it is not a firm requirement to apply for the classical equations, it still can be helpful. Now, let us recall the Guyer–Krumhansl equation again as a PDE system:

$$\rho c \partial_t T + \nabla \cdot \mathbf{q} = 0,$$
$$\tau \partial_t \mathbf{q} + \mathbf{q} = -\lambda \nabla T + l^2 \Delta \mathbf{q}, \tag{21}$$

with heat pulse boundary condition. Then, putting the field variables at the same grid point would require a compatible definition for the temperature on the boundary, too. As it is not possible, we propose a different method, constructing a staggered grid, as the Fig. 1 shows for a one-dimensional situation. It avoids defining the temperature on the boundary by distinguishing 'surface' and 'volume' quantities. In this particular setting, the heat flux is situated on the boundary of each cell, and the temperature characterizes their volume average. This is a sort of finite volume approach. That choice is made arbitrarily, it is also possible to define the temperature on the boundary, but it is not natural from a physical point of view.

More importantly, such a staggered discretization is strongly related to the structure of the equations that is restricted by the first and the second law of thermodynamics. Thus it remains applicable for any PDE, at least in the linear regime, which respects the structure and compatibility with thermodynamics.

As an example, we demonstrate this numerical approach on the one-dimensional version of the Guyer–Krumhansl model. We start with discretization explicitly the balance equation,

$$T_j^{n+1} = T_j^n - \frac{\Delta t}{\rho c \Delta x}(q_{j+1}^n - q_j^n), \tag{22}$$

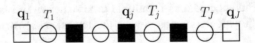

Fig. 1 The 1D discretization of the Guyer–Krumhansl equation

and continue with the constitutive equation:

$$q_j^{n+1} = q_j^n - \frac{\Delta t}{\tau} q_j^n - \frac{\lambda \Delta t}{\tau \Delta x}(T_j^n - T_{j-1}^n) + \frac{l^2 \Delta t}{\tau \Delta x^2}(q_{j+1}^n - 2q_j^n + q_{j-1}^n), \quad (23)$$

where the forward approximation for the time derivatives are used, and discretizing the spatial derivatives according to the staggered field. As always the lowest order term restricts the precision, it becomes a first order scheme. However, it still remains an efficient way to obtain a physically admissible solution. The temperature field is calculated as a consequence of the heat flux, exactly as the analytical solution suggests, without using any boundary condition for T.

This scheme is improved to be second order with taking the convex combination of the explicit and the implicit terms and formulating a Crank–Nicolson-like scheme (which acts analogously as the trapezoidal integration) [1, 28]:

$$\rho c \frac{1}{\Delta t}(T_j^{n+1} - T_j^n) = -\frac{1}{\Delta x}\left[(1 - \Theta)\left(q_{j+1}^n - q_j^n\right) + \Theta\left(q_{j+1}^{n+1} - q_j^{n+1}\right)\right], \quad (24)$$

and for the constitutive equation, too:

$$\frac{\tau}{\Delta t}\left(q_j^{n+1} - q_j^n\right) + \left[(1 - \Theta)q_j^n + \Theta q_j^{n+1}\right]$$

$$+ \frac{\lambda}{\Delta x}\left[(1 - \Theta)(T_j^n - T_{j-1}^n) + \Theta(T_j^{n+1} - T_{j-1}^{n+1})\right] \quad (25)$$

$$- \frac{l^2}{\Delta x^2}\left[(1 - \Theta)(q_{j+1}^n - 2q_j^n + q_{j-1}^n) + \Theta(q_{j+1}^{n+1} - 2q_j^{n+1} + q_{j-1}^{n+1})\right] = 0,$$

where Θ denotes the combination parameter and $\Theta = 1/2$ leads to a second order scheme. The reasoning behind that idea is simple: the first order derivatives (both time and space) are not situated at the same grid point, see Fig. 2 wherein h denotes an arbitrary index, could be either related to time or space. In case of the time derivatives, the derivative itself is situated between two levels of n and $n + 1$. Hence the convex combination of the other terms with $\Theta = 1/2$ puts them onto the same level. Moreover, that idea also works for the spatial derivatives on the staggered field, but no linear combination is needed.

Fig. 2 The place of the derivatives on the lattice

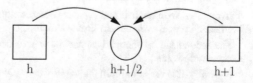

h   h+1/2   h+1

4.3 Stability Analysis

So far, we already mentioned a few words about 'stability' as some property of the numerical approach. We have to distinguish physical and numerical stabilities since they are completely different notions. The physical stability corresponds to the physical content of the PDE, without any approximation. The numerical stability, however, is only a property which is introduced by the approximations. Considering a particular PDE with a particular numerical approximation, it is possible to investigate the resulting stability properties.

Since the compatibility with thermodynamics is a strong requirement for mathematical models, the resulting constitutive equations (together with the corresponding balances) have asymptotically stable[22] solutions.[23] However, it is not difficult to obtain numerically unstable solutions for physically stable processes.

The numerical stability—instability is in close connection to the boundedness of the errors, introduced due to the approximations. A scheme remains stable if the errors are bounded. As a thumb rule, the explicit schemes—those that use values from the previous time instants only—have a stability condition, independently of their precision.[24] The implicit schemes have much better stability attributes.

Now we continue the example on the Guyer–Krumhansl equation with investigating the stability properties. In most cases, the von Neumann and Jury methods are adequate for linear stability analysis [1, 35]. According to the von Neumann method, let us assume the solution of the difference Eqs. (24) and (25) in the form[25]

$$\phi_j^n = \xi^n e^{ikj\Delta x}, \tag{26}$$

where i is the imaginary unit, k is the wave number, and $j\Delta x$ denotes the jth spatial step, and ξ is called the growth factor[26] [1]. The scheme is stable if and only if $|\xi| \leq 1$ holds, this inequality serves as the stability criteria. Using (26) one can express each term from (24) to (25), for example $q_{j-1}^{n+1} = \xi^{n+1} e^{ik(j-1)\Delta x} \cdot q_0$. Substituting (26) into (24) and (25) yields

$$T_0(\xi - 1) + q_0 \frac{\Delta t}{\rho c \Delta x} \left[(1 - \Theta)\left(e^{ik\Delta x} - 1\right) + \Theta\xi\left(e^{ik\Delta x} - 1\right) \right] = 0, \tag{27}$$

---

[22]It requires an attractive and Lyapunov-stable equilibrium. The simplest example is a cooling/heating process without any heat generation.

[23]The second law of thermodynamics itself is a stability theory, that is, the total entropy is a Lyapunov function [34].

[24]Naturally, it is easier to fulfill the stability conditions for schemes with higher precision.

[25]Realize that ξ is analogous to $e^{i\omega\Delta t}$, and this stability analysis uses the discrete version of dispersion relations. In fact, ξ itself is the dispersion relation.

[26]It is a complex number, in general.

$$q_0(\xi - 1) + q_0 \frac{\Delta t}{\tau_q} [(1 - \Theta) + \Theta\xi]$$

$$+ T_0 \frac{\lambda \Delta t}{\Delta x \tau_q} [(1 - \Theta)(1 - e^{-ik\Delta x}) + \Theta\xi(1 - e^{-ik\Delta x})]$$

$$- q_0 \frac{l^2 \Delta t}{\Delta x^2 \tau_q} [(1 - \Theta)(e^{ik\Delta x} - 2 + e^{-ik\Delta x}) + \Theta\xi(e^{ik\Delta x} - 2 + e^{-ik\Delta x})] = 0.$$

$$(28)$$

Now, one constructs a coefficient matrix $\mathbf{M}$ such as $\mathbf{M} \cdot \mathbf{u} = 0$, with $\mathbf{u}^T = [T_0; q_0]$. Calculating the determinant of $\mathbf{M}$ leads to the characteristic polynomial of system (27) in the form $F(\xi) = a_2 \xi^2 + a_1 \xi + a_0$ with the coefficients, leaving the choice free for Θ:

$$a_0 = 1 - \frac{\Delta t}{\tau_q}(1 - \Theta) + [2\cos(k\Delta x) - 2](1 - \Theta)\frac{\Delta t}{\Delta x^2 \tau_q} [l^2 - \Delta t(1 - \Theta)],$$

$$a_1 = -2 + \frac{\Delta t}{\tau_q}(1 - 2\Theta) + [2\cos(k\Delta x) - 2]\frac{\Delta t}{\Delta x^2 \tau_q} [l^2(2\Theta - 1) - 2\Delta t(1 - \Theta)\Theta],$$

$$a_2 = 1 + \frac{\Delta t}{\tau_q}\Theta - [2\cos(k\Delta x) - 2]\frac{\Delta t}{\Delta x^2 \tau_q}\Theta(l^2 + \Delta t\Theta). \qquad (29)$$

Our aim is to restrict the coefficients in a way which ensures that the roots of the present polynomial stay within the unit circle in the complex plane. That roots are the eigenvalues of matrix $\mathbf{M}$. These restrictions can be derived using the so-called Jury criteria.[27] At this level, the stability condition $|\xi| \leq 1$ means an inequality between the time and space steps. In general, the time step Δt is constrained as a function of Δx. Nevertheless, the material parameters also appear in these relations and this is how the scaling properties of the mathematical model influence the stability. There are $m + 1$ criteria for an mth order polynomial:

1. $F(\xi = 1) \geq 0$,
2. $F(\xi = -1) \geq 0$,
3. $a_2 \geq |a_0|$.

It is possible to express each condition also as a function of Θ. However, this would not be useful for practical purposes and now we consider the $\Theta = 1$ case, that gives

1. $\frac{4\Delta t^2}{\tau_q \Delta x^2} > 0$,
2. $4 + 2\frac{\Delta t}{\tau_q} + 4\frac{\Delta t}{\tau_q \Delta x^2}(2l^2 + \Delta t) > 0$,
3. $1 < 1 + \frac{\Delta t}{\tau_q} + 4\frac{\Delta t}{\tau_q \Delta x^2}(l^2 + \Delta t)$,

hence, the scheme remains unconditionally stable as long as all parameters are positive, i.e., satisfying the thermodynamical conditions, too. The $\Theta = 1$ stands for the completely implicit approach which, as it is proved here, requires no condition

---

[27]The Routh–Hurwitz criteria are applicable for continuous systems.

between the Δt and Δx. However, it is only a first order scheme and the more interesting case belongs to $\Theta = 1/2$, in which the precision is increased. Repeating the
same calculation yields

1. $\frac{4\Delta t^2}{\tau_q \Delta x^2} > 0$,
2. $4 > 0$,
3. $0 < 1 + \frac{4l^2}{\Delta x^2}$,

that is, the scheme is still unconditionally stable. It could be surprising as there
are explicit parts in the Eqs. (24) and (25) as well. In our last remark, we note that
the present scheme is also applicable for the submodels of the Guyer–Krumhansl
equation, i.e., for $l = 0$, each criterion are fulfilled. Therefore, the stability is proved.

4.4 Discretization of Mechanical Systems

The essential aspects remain the same for mechanical systems, such as the treatment
of boundary conditions, being aware of the time scales (or wave propagation speeds,
accordingly) and preserving the physical content while the scheme converges to a
real, asymptotically stable solution.

One essential difference between the thermal and mechanical problems originates
in physics, naturally: the simplest mechanical model—the Hooke's law—describes
a completely elastic behavior. Elasticity does not produce entropy, and thus it is a
reversible process in which the mechanical (kinetic and elastic) energy is conserved.
Usually, in the engineering handbooks, such physical phenomenon is modeled with
a wave equation for the velocity field $\mathbf{v}$,

$$\partial_{tt}\mathbf{v} = v_0 \Delta \mathbf{v}, \tag{30}$$

where v_0 is the characteristic propagation speed. Unfortunately, that approach makes
difficult to notice the Hamiltionian dynamics (energy conservation). Likewise to the
thermal problems, we propose to solve the PDEs separately, that is,

$$\rho \partial_t \mathbf{v} + \nabla \cdot \mathbf{P} = 0, \tag{31}$$

$$-\mathbf{P} = \sigma = 2\mu\varepsilon + \lambda \varepsilon^{sph}, \tag{32}$$

$$\partial_t \varepsilon = (\nabla \mathbf{v})_{\text{sym}}, \tag{33}$$

where Eq. (31) describes the momentum balance with the pressure tensor $\mathbf{P}$, and
$\nabla \cdot \mathbf{P}$ is the force acting locally. Equation (32) is the constitutive relation between
the pressure and the deformation ε, including the Lamè coefficients μ and λ. The
system (31)–(33) is only valid for small deformations. The Hamiltionian structure
becomes apparent in 1D:

$$\rho \partial_t v = E \partial_x \varepsilon,$$

$$\partial_t \varepsilon = \partial_x v.$$

Fig. 3 The 1D discretization of the elastic (Hamiltonian) mechanical system

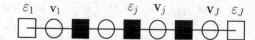

That form now reflects the usual structure of Hamiltion equations, and hence the energy conservation becomes visible, too.

First, we remark that the same spatial discretization method of staggered fields is applicable, it is our next example. Its advantage becomes exceptionally significant for rheological equations in which the constitutive relation becomes a differential equation, and still useful even in the classical case. Therefore, the two field variables are situated on different grid points, it is depicted in Fig. 3. In this simple one-dimensional example, using the deformation as a primary variable is equivalent to the stress due to the constant proportionality.

We draw attention again to the energy conservation, that property must be included in the numerical scheme as well, i.e., the time-stepping algorithm of the discretized system respects the energy conservation. Such a time-stepping method is called symplectic integrator [36–38] and can be combined with the staggered field approach. This is the most efficient way to fulfill the previous requirement. The simplest, first-order symplectic integration is a so-called semi-implicit Euler method, that is,

$$\varepsilon_j^{n+1} = \varepsilon_j^n + \frac{\Delta t}{\Delta x}\left(v_j^n - v_{j-1}^n\right), \tag{34}$$

$$v_j^{n+1} = v_j^n + \frac{E\Delta t}{\rho\Delta x}\left(\varepsilon_{j+1}^{n+1} - \varepsilon_j^{n+1}\right). \tag{35}$$

While Eq. (34) is explicit, the other one is implicit, and both of them use forward differencing in time. However, in Eq. (35), the new values of ε^{n+1} appear, instead of ε^n. This is the key step to make the time-stepping symplectic. Since the first time step for a variable is always explicit, there is a condition for stability.

Stability Analysis

Using again the von Neumann and Jury methods and repeating the same steps, that is, assuming the solution of the difference equation in the following form,

$$\varphi_j^n = \xi^n e^{ikj\Delta x}, \tag{36}$$

and omitting the detailed derivation, the characteristic polynomial reads

$$F(\xi) = \xi^2 - \xi\left(2 + \frac{\Delta t^2}{\Delta x^2}\frac{E}{\rho}\left(2\cos(k\Delta x) - 2\right)\right) + 1, \tag{37}$$

that can be solved in some cases, for instance, the solution is

$$\xi = \frac{1}{\Delta x^2 \rho}\Big(\Delta x^2 \rho + \Delta t^2 E\big(-1 + \cos(\Delta xk)\big)\pm$$
$$\sqrt{\Delta t^2 E(-1 + \cos(\Delta xk))\big(-\Delta t^2 E + 2\Delta x^2 \rho + \Delta t^2 E \cos(\Delta xk)\big)}\Big)$$

(38)

and applying directly the stability requirement $|\xi| \leq 1$, would lead to the stability conditions. However, it is easier to apply the Jury criteria which yield

- The $F(\xi = 1) \geq 0$ condition is trivially fulfilled: $F(\xi = 1) = 4\frac{\Delta t^2}{\Delta x^2}\frac{E}{\rho} > 0$.
- The $F(\xi = -1) \geq 0$ condition leads to the CFL stability criterion: $\frac{\Delta x}{\Delta t} \geq v_0$, where $v_0 = \sqrt{\frac{E}{\rho}}$.
- The last one is: $|a_2| \geq |a_0|$, that relation is automatically fulfilled because both of them is 1.

From the viewpoint of time and spatial scales, it is remarkable how the stability condition restricts the time or the space step. If one of them is fixed (usually, the spatial part), then the other one must be in agreement with the stability condition. In general, it requires extremely small time steps. For instance, using $\Delta x = 0.01$ m, for v_0 being 5000 m/s, then $\Delta t \leq 2 \cdot 10^{-6}$ s. Instead, it is worth to use dimensionless parameters by introducing a dimensionless velocity such as $\hat{v} = v/v_0$. Moreover, using this characteristic velocity, and a unit length L, $\hat{x} = x/L$, $\hat{t} = tv_0/L$ with $\hat{\sigma} = \sigma/(\rho v_0^2)$, then it results in a more convenient stability condition: $\Delta\hat{x} \geq \Delta\hat{t}$. For further examples of introducing dimensionless quantities and the related mathematical notions, we refer to [39]; the corresponding section can be found in Appendix A.[28]

Applying this simple symplectic scheme, it becomes possible to model a one-dimensional, purely elastic wave propagation, see Fig. 4 for the demonstration. In order to preserve that characteristic of the solution, it must be free from dissipative and dispersive errors that jeopardize the complete calculation. Due to their importance, we discuss these sources of numerical errors in the following.

4.5 Numerical Errors: Dispersion or Dissipation?

Keeping the finite difference method in our focus, we point out again that it provides only approximative solutions to the differential equations. One source of the numerical error roots exactly in the truncation error of the Taylor series expansion. We call it 'inaccuracy' as the value of the solution differs from the exact one, but the numerical solution still represents the characteristics adequately. Unfortunately, this is not the only consequence of the approximation, it introduces the artificial phenomena of numerical dissipation and dispersion.

At this point, we have to highlight some remarks about the following discussion. There is no clear mathematical definition for these numerical errors. On the one hand,

---

[28] Taken in accordance with the Creative Commons License.

the derivation of ξ is the discrete version of dispersion relation (i.e., the function $\omega = \omega(k)$), and thus it inherits all the problems as well. For example, the usual derivation of the relationship $\omega(k)$ assumes a plane-wave solution, with arbitrarily assuming that the frequency ω is complex and the wavenumber k is real or vice versa. Moreover, that approach completely omits the boundary conditions despite they affect the definition of k since the solution is represented in the eigensystem $\exp(ikj\Delta x)$. Hence k cannot be arbitrary, and the boundary effects should not be neglected. However, so far, these aspects are not elaborated in regard to dispersion relations.

Consequently, the stability conditions based on $|\xi| \leq 1$ can be slightly inaccurate in some cases. In the following analysis, we show a few estimations about the dissipation and dispersion using ξ. Furthermore, we also present the boundary effect on the mentioned numerical errors.

Dissipation

Usually, the dissipation error is illustrated on a sharp edge, which is smeared out slowly, resulting in a false solution. In other words, the amplitude of the wave is decreasing in every time step and occurs when $|\xi| < 1$. It stabilizes the numerical solution. For $|\xi| = 1$, the scheme can be called 'conservative'; only when no dispersive error is present. This is exactly the requirement for symplectic methods. For instance, applying any Δt and Δx in accordance with the stability condition, Eq. (38) satisfies $|\xi| = 1$ (Fig. 5). Moreover, to obtain a solution such as in Fig. 4, also requires $v_0 \Delta t = \Delta x$ in order to vanish the dispersion effects. Thus the numerical dissipation and dispersion errors are not independent of each other.

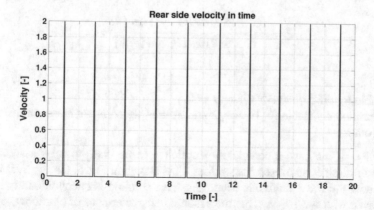

Fig. 4 Rear side velocity in time, using dimensionless quantities. The boundary condition is defined for ε, the excitation is applied similarly to heat conduction, using Eq. (10)

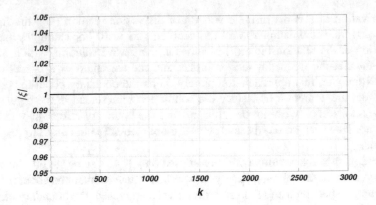

Fig. 5 Equation (38) for $|\xi|$ as a function of k

However, the presence of dissipation and dispersion is a sort of characteristic property of a scheme and varies from scheme to scheme. For comparative reasons, let us demonstrate that property using the following possibility for discretization,

$$\varepsilon_j^{n+1} = \varepsilon_j^n + \Theta \frac{\Delta t}{\Delta x}\left(v_j^n - v_{j-1}^n\right) + (1 - \Theta)\frac{\Delta t}{\Delta x}\left(v_j^{n+1} - v_{j-1}^{n+1}\right), \tag{39}$$

$$v_j^{n+1} = v_j^n - \Theta \frac{E\,\Delta t}{\rho\Delta x}\left(\varepsilon_{j+1}^n - \varepsilon_j^n\right) - (1 - \Theta)\frac{E\,\Delta t}{\rho\Delta x}\left(\varepsilon_{j+1}^{n+1} - \varepsilon_j^{n+1}\right), \tag{40}$$

which is similar to the one applied for heat conduction, and for $\Theta = 1/2$, it is supposed to be second order accurate. Repeating the stability analysis, one can obtain an expression for ξ:

$$\xi = \frac{\Delta x \pm 2iv_0\Delta t(\Theta - 1)\sin\left(\frac{\Delta xk}{2}\right)}{\Delta x \pm 2iv_0\Delta t\Theta \sin\left(\frac{\Delta xk}{2}\right)}, \tag{41}$$

which is

- unstable for all Δt and Δx when $\Theta = 0$,
- unconditionally stable for $\Theta = 1/2$ and 1, thus we compare these parameters to each other.

When $\Theta = 1$, we have a completely implicit scheme that produces a significant amount of dissipation, the corresponding $|\xi|$ is shown in Fig. 6. On Fig. 7, it is presented how the wave amplitude decreases. This solution is unconditionally stable, works even with $v_0\Delta t > \Delta x$, but the essential characteristics of elastic wave propagation is completely distorted due to the dissipation error.

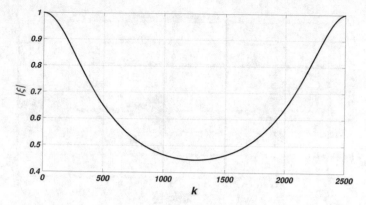

Fig. 6 Equation (41) with $\Theta = 1$, showing $|\xi|$ as a function of k, with normalizing the velocity by v_0

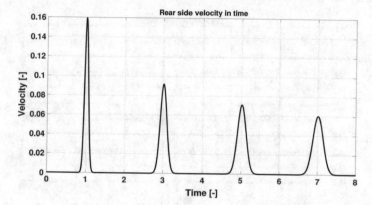

Fig. 7 One dissipative solution for elastic wave propagation using the schemes in (39) and (40)

Dispersion

We continue our comparative analysis here since the dispersion error becomes apparent now, for both schemes, considering symplectic time-stepping or $\Theta = 1/2$ combination. The dispersion error occurs as a spurious oscillation around sharp edges. Usually, it distorts the solution in a way which makes the simulation meaningless. Its source is coded into ξ as well; however, its investigation is not so 'algorithmic'.

The dispersion error is often referred to be as 'phase error', it will be visible when ξ is depicted on the complex plane as a function of k. In general, it is investigated with numerical simulations, trying to reconstruct the exact dispersion relation. The error can be estimated as a difference among them. For the presented symplectic Euler method, dispersion error occurs when $v_0 \Delta t < \Delta x$, and the solution is shown in Fig. 8. At the same time, that figure also shows a dissipative error, they appear together.

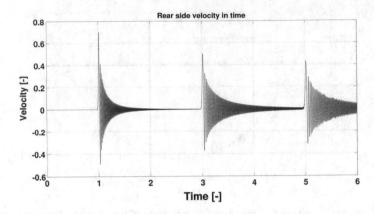

Fig. 8 One dispersive (and dissipative) solution for elastic wave propagation using the symplectic Euler scheme with $\Delta t = 0.95 \Delta x$

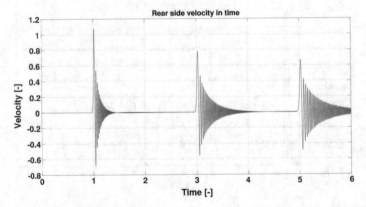

Fig. 9 One dispersive (and dissipative) solution for elastic wave propagation using the $\Theta = 1/2$ scheme

We note that for the following test simulations, we use dimensionless parameters (i.e., $v_0 = 1$) such as $\Delta x = 0.002$, and the mentioned 'pulse-like' excitation is $t_p = 0.01$.

Before using the $\Theta = 1/2$ scheme, we recall that it is unconditionally stable, i.e., it remains stable for any Δt. Although $|\xi| = 1$ holds for any Δt and Δx, it still produces a dissipative and dispersive error, that is, $|\xi| = 1$ is a necessary but not satisfactory requirement to being symplectic. For instance, applying the same spatial and time steps as previously, we obtain a solution, shown in Fig. 9. It is not a meaningful solution due to dispersive-dissipative errors. Thus we emphasize that a scheme remains conservative only when $|\xi| = 1$ and no phase error is occurring. Plotting ξ on the complex plane for both schemes, the differences become apparent.

Symplectic scheme: Beginning with the symplectic scheme, we observe that the conservative solution requires $|\xi| = 1$ for all k. Moreover, it must consist of the

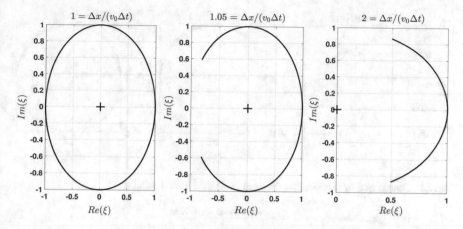

Fig. 10 Change of ξ with decreasing the time step, using the symplectic Euler scheme

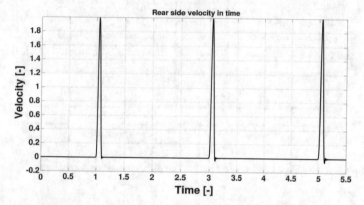

Fig. 11 The effect of changing t_p to 0.1, the dispersive error is significantly reduced, using the symplectic Euler scheme with $\Delta t = 0.95 \Delta x$

complete set of points on the unit circle. It means that the scheme itself is not sensitive to the boundary condition (e.g., to the 'pulse' length t_p), and the conservation property is not restricted to a particular wavenumber[29] k. When the relation $v_0 \Delta t < \Delta x$ holds, that 'completeness' is violated and the dispersive error is starting to appear for certain k (Fig. 10). However, increasing[30] t_p from 0.01 to 0.1 shows that the dispersive (and the dissipation, accordingly) error is decreased (Fig. 11). In other words, the scheme becomes sensitive to certain wavenumbers, and that behavior depends on the boundary conditions.

Θ-**schemes**: For $\Theta = 1$, the dissipation always over-dominates the dispersion, and it is not possible to approach the conservative solutions (Fig. 12). Using $\Delta x / \Delta t = 20$,

---

[29] Naturally, for a reasonable numerical resolution.

[30] That is, modifying the excitation in the boundary condition.

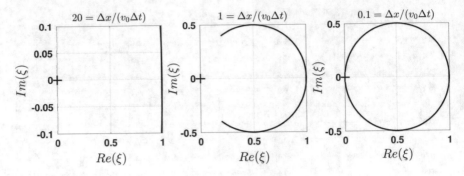

Fig. 12 Change of ξ by changing the time step, using the $\Theta = 1$ scheme

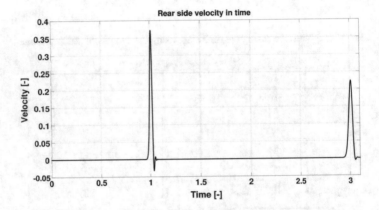

Fig. 13 Solution with $\Theta = 1$ and $\Delta x / \Delta t = 20$. The dispersive error is apparent

the dispersive error becomes visible, but changing 20 to 1, it completely disappears and investigating the wave amplitude, the dissipation error is significantly increased (Fig. 13). Moreover, changing t_p from 0.01 to 0.1, the dissipative error is also decreased, hence the appearance of the numerical errors also depends on the boundaries (Fig. 14). That dependence on t_p is also visible for $\Theta = 1/2$ scheme, comparing Fig. 9 to Fig. 15. That scheme can approximate the symplectic Euler method with a large Δt (Fig. 16).

5   Outlook and Summary

We have seen that the symplectic schemes are the best choice to simulating conservative systems as even the simplest first-order accurate Euler method offers a fast and correct solution. Moreover, when $v_0 \Delta t = \Delta x$, the dissipation and dispersion errors remain independent of the excitation frequency. This does not hold to the Θ-methods.

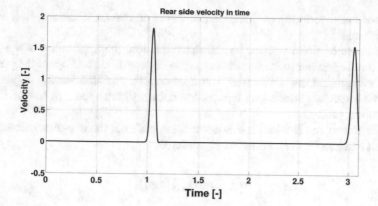

Fig. 14 Changing t_p to 0.1 with $\Delta x/\Delta t = 20$ and $\Theta = 1$

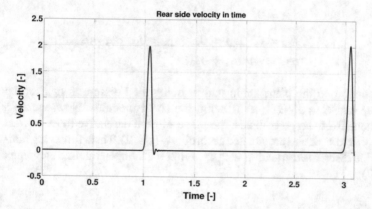

Fig. 15 Changing t_p to 0.1 with $\Delta x/\Delta t = 1$ and $\Theta = 1/2$

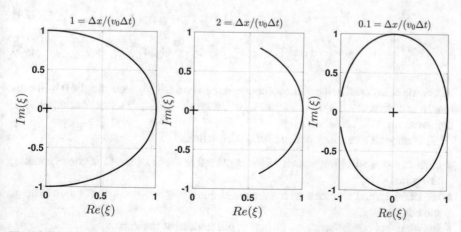

Fig. 16 Change of ξ by changing the time step, using the $\Theta = 1/2$ scheme

They are sensitive to the parameter t_p; however, in certain situations, they offer a good alternative.

The common point was the spatial discretization, the application of staggered fields. It makes easier to implement any boundary condition and opens the possibility for generalized models as well. Here, its one-dimensional version is presented only, but it can be extended for tensorial variables which appear in two and three-dimensional problems.

For instance, the Hooke's law is not that simple as in the one-dimensional case, for each component in two dimension, we have

$$\varepsilon_{11} = \frac{1}{E}\big(\sigma_{11} - \nu(\sigma_{22} + \sigma_{33})\big),$$

$$\varepsilon_{22} = \frac{1}{E}\big(\sigma_{22} - \nu(\sigma_{11} + \sigma_{33})\big),$$

$$\varepsilon_{33} = 0, \quad \varepsilon_{12} = \frac{\sigma_{12}}{2G}, \tag{42}$$

$$\varepsilon_{13} = 0, \quad \varepsilon_{23} = 0, \quad \sigma_{13} = 0, \quad \sigma_{23} = 0,$$

$$\sigma_{33} = -\nu(\sigma_{11} + \sigma_{22}),$$

corresponding to the plane strain with ν being the Poisson ratio, E is the Young modulus and $G = E/(2(1 + \nu))$ being the shear modulus. Now, we refer to the fundamentals of staggered fields: the place of each derivative must be tracked and the basic field variables are situated accordingly (Fig. 2). Then, following that way, we obtain the discretized equation with applying the symplectic time stepping strategy (Fig. 17):

$$_{11}\varepsilon_{i,j}^{n+1} = _{11}\varepsilon_{i,j}^{n} + \frac{\Delta t}{\Delta x}(_x v_{i+1,j}^n - _x v_{i,j}^n), \tag{43}$$

$$_{12}\varepsilon_{i,j}^{n+1} = _{12}\varepsilon_{i,j}^{n} + \frac{\Delta t}{2\Delta y}(_x v_{i,j+1}^n - _x v_{i,j}^n) + \frac{\Delta t}{2\Delta x}(_y v_{i+1,j}^n - _y v_{i,j}^n), \tag{44}$$

$$_x v_{i,j}^{n+1} = _x v_{i,j}^{n} + \frac{\Delta t}{\rho \Delta x}\big(_{11}\sigma_{i,j}^{n+1} - _{11}\sigma_{i-1,j}^{n+1}\big) + \frac{\Delta t}{\rho \Delta y}\big(_{12}\sigma_{i,j}^{n+1} - _{12}\sigma_{i-1,j}^{n+1}\big), \tag{45}$$

where the equations for the other components can be derived, accordingly. It is important to note that σ^{n+1} is used instead of σ^n, after determining all ε^{n+1} components, the stress must be calculated.

That approach can be extended for 3D equations:

- each component of the velocity field is placed at the middle of the corresponding ortogonal side,
- the off-diagonal elements of the tensorial quantities are situated at the middle of the edges,
- the diagonal components are placed at the middle of the cube.

Fig. 17 The discretization of the mechanical system for 2D elastic wave propagation

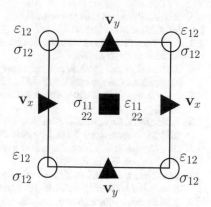

As a final note, the discretization of the ballistic-conductive equation is analogous with the substitution: $\varepsilon \to \mathbf{Q}$, $\mathbf{v} \to \mathbf{q}$; and the temperature (as a scalar) is situated at the center of each cell.

Overall, we want to emphasize that a numerical method is chosen to the actual physical content of the equations. While the symplectic methods are much better applicable for conservative systems, their advantage disappears for (physically) dissipative situations and leaves space for other methods. Moreover, it is demonstrated how easy to obtain unphysical but stable numerical solutions. Thus it is strongly recommended to pay attention to any numerical methods, and take any opportunity to validate them. For further examples, including rheological models, we refer to [40].

References

1. W.H. Press, *Numerical Recipes 3rd Edition: The Art of Scientific Computing* (Cambridge University Press, 2007)
2. L.C. Evans. Partial Differential Equations. *Graduate Studies in Mathematics* (1997)
3. P.K. Kythe, M.R. Schäferkotter, P. Puri, *Partial Differential Equations and Mathematica* (CRC Press, Boca Raton, 2002)
4. S.J. Farlow, *Partial differential equations for scientists and engineers* (Courier Corporation, North Chelmsford, 1993)
5. ANSYS Inc. ANSYS Fluent Theory Guide 2019 R2 (2019)
6. M.M. Woolfson, G.J. Pert, *An Introduction to Computer Simulation* (Oxford University Press on Demand, Oxford, 1999)
7. R. Kovács, Analytic solution of Guyer-Krumhansl equation for laser flash experiments. Int. J. Heat Mass Transf. **127**, 631–636 (2018)
8. M. Necati Ozisik, *Heat Conduction* (Wiley, New York, 1993)
9. H.S. Carslaw, J.C. Jaeger, *Conduction of heat in solids* (1959)
10. Gregory T. von Nessi, *Analytic Methods in Partial Differential Equations* (Springer, Berlin, 2011)
11. K. Zhukovsky, Violation of the maximum principle and negative solutions for pulse propagation in Guyer-Krumhansl model. Int. J. Heat Mass Transf. **98**, 523–529 (2016)

12. K.V. Zhukovsky, Exact solution of Guyer-Krumhansl type heat equation by operational method. Int. J. Heat Mass Transf. **96**, 132–144 (2016)
13. K.V. Zhukovsky, Operational approach and solutions of hyperbolic heat conduction equations. Axioms **5**(4), 28 (2016)
14. K.V. Zhukovsky, H.M. Srivastava, Analytical solutions for heat diffusion beyond Fourier law. Appl. Math. Comput. **293**, 423–437 (2017)
15. S.C. Chapra, R.P. Canale, *Numerical methods for engineers*, vol. 2 (McGraw-Hill, New York, 1998)
16. L. Rezzolla, *Numerical Methods for the Solution of Hyperbolic Partial Differential Equations* (SISSA, International School for Advanced Studies, Trieste, 2005)
17. P.G. Ciarlet, J.-L. Lions, *Handbook of Numerical Analysis: Finite Element Methods (Part 1)*, vol. 2 (North Holland, 1991)
18. P.G. Ciarlet, J.-L. Lions, *Handbook of Numerical Analysis: Finite Eelement Methods (Part 2), Numerical Methods for Solids (Part 2)*, vol. 4 (North Holland, 1996)
19. S. Moaveni, *Finite Element Analysis - Theory and Application with ANSYS* (Prentice Hall, New Jersey, 1999)
20. D.T. Bailey, *Meteorological monitoring guidance for regulatory modeling applications* (DIANE Publishing, Collingdale, 2000)
21. M.T. Manzari, M.T. Manzari, On numerical solution of hyperbolic heat conduction. Commun. Numer. Methods Eng. **15**(12), 853–866 (1999)
22. B. Xu, B.Q. Li, Finite element solution of non-fourier thermal wave problems. Numer. Heat Transf. Part B Fundam. **44**(1), 45–60 (2003)
23. S. Bargmann, P. Steinmann, Finite element approaches to non-classical heat conduction in solids. Comput. Model. Eng. Sci. **9**(2), 133–150 (2005)
24. S. Bargmann, P. Steinmann, Modeling and simulation of first and second sound in solids. Int. J. Solids Struct. **45**(24), 6067–6073 (2008)
25. H. Rahideh, P. Malekzadeh, M.R.G. Haghighi, Heat conduction analysis of multi-layered FGMs considering the finite heat wave speed. Energy Convers. Manag. **55**, 14–19 (2012)
26. V. Vishwakarma, A.K. Das, P.K. Das, Analysis of non-Fourier heat conduction using smoothed particle hydrodynamics. Appl. Therm. Eng. **31**(14–15), 2963–2970 (2011)
27. S. Bargmann, A. Favata, Continuum mechanical modeling of laser-pulsed heating in polycrystals: A multi-physics problem of coupling diffusion, mechanics, and thermal waves. ZAMM-Journal of Applied Mathematics and Mechanics/Zeitschrift für Angewandte Mathematik und Mechanik **94**(6), 487–498 (2014)
28. Á. Rieth, R. Kovács, T. Fülöp, Implicit numerical schemes for generalized heat conduction equations. Int. J. Heat Mass Transf. **126**, 1177–1182 (2018)
29. R. Kovács, Heat conduction beyond Fourier's law: theoretical predictions and experimental validation. Ph.D. Thesis, Budapest University of Technology and Economics (BME), 2017
30. P.D. Lax, R.D. Richtmyer, Survey of the stability of linear finite difference equations. Commun. Pure Appl. Math. **9**(2), 267–293 (1956)
31. P.D. Lax, *Hyperbolic Partial Differential Equations* (American Mathematical Society, Providence, 2006)
32. D.R. Durran, The third-order Adams-Bashforth method: An attractive alternative to leapfrog time differencing. Mon. Weather. Rev. **119**(3), 702–720 (1991)
33. J.C. Butcher, *The numerical analysis of ordinary differential equations: Runge-Kutta and general linear methods*, vol. 512 (Wiley, New York, 1987)
34. T. Matolcsi, *Ordinary Thermodynamics* (Akadémiai Kiadó, Budapest, 2004)
35. E.I. Jury, *Inners and Stability of Dynamic systems* (Wiley, London, 1974)
36. T.S. Biró, Conserving algorithms for real-time nonabelian lattice gauge theories. Int. J. Mod. Phys. C **6**(03), 327–344 (1995)
37. E. Hairer, C. Lubich, G. Wanner, Geometric numerical integration illustrated by the störmer-Verlet method. Acta Numer. **12**, 399–450 (2003)
38. H.C. Öttinger, GENERIC integrators: structure preserving time integration for thermodynamic systems. J. Non Equilib. Thermodyn. **43**(2), 89–100 (2018)

39. T. Fülöp. Chapters in thermodynamics. http://energia.bme.hu/~fulop/UVHT/ (2019). [Online; Accessed 14 Aug 2019]
40. T. Fülöp, R. Kovács, M. Szücs, M. Fawaier. Thermodynamically extended symplectic numerical scheme with half space and time shift applied for rheological waves in solids (2019). Submitted, arXiv:1908.07975

Appendix A
Switch to Dimensionless Quantities—In General and for VdW

Similarly to how, in fluid mechanics, rescaling quantities by constants to make them dimensionless and forming dimensionless combinations of constants are beneficial for various purposes—realizing similarity among different setups and establishing the appropriate rescaling of one setup to another, finding a universal common behaviour of systems with different sizes, and reducing the number of free parameters to the *essentially independent/free* parameters for reducing the dimension of the parameter space to scan for numerical calculations and diagrams—, the effort to reduce the number of free parameters and to have dimensionless quantities is similarly advantageous in thermodynamics. Before considering the example of the Van der Waals gas, let us summarize what dimensionful quantities are and how to practically treat them.

When we walk in a forest, we can find sticks and rods of various size. Similarly, at home we find pens, pencils, pieces of thread etc. of various size. Then we gradually invent the abstraction of *the set of possible length values*. We find that lengths can be compared whether they are of the same lengths or one is larger than the other. We can add lengths (one stick as the continuation of the other), subtract them (backward continuation), multiply them by numbers,[1] with natural operation rules (associativity, distributivity etc.). Altogether, we find that the set of possible length values, denoted here by $\mathbb{L}$, is a one-dimensional real oriented vector space.[2] Hereafter, a one-dimensional real oriented vector space is called a *measure line*, which is a much shorter and friendlier name.[3]

---

[1] Multiplication by integers: repeating a length n times; division by integers: folding a piece of thread into two, three etc.; multiplication by rational numbers: a multiplication and a division; irrational numbers: as a limit of rational ones.

[2] Initially, the set of possible length values contains only positive values but, since we frequently find it convenient to use space coordinates like $x = -3.2\,\text{m}$, we extend it to negative values, too. 'Oriented' means here that we have distinguished one half of the vector space as positive, the other half being the negative one.

[3] A line used for quantities, that is, for measurement purposes.

© Springer Nature Switzerland AG 2020
V. Józsa and R. Kovács, *Solving Problems in Thermal Engineering*, Power Systems, https://doi.org/10.1007/978-3-030-33475-8

For the possible time interval values, we introduce another measure line, denoted by $\mathbb{T}$. For the possible mass values, yet another measure line, $\mathbb{M}$ is needed;[4] and we will use a fourth measure line, $\mathbb{H}$, for temperature values.[5] Adding a length l and a time interval t is meaningless (physically, as well as mathematically you cannot add elements coming from different sets). However, their product and their quotient are meaningful—mathematically, the product lt lives in the tensorial product of $\mathbb{L}$ and $\mathbb{T}$, $\mathbb{LT}$, which is also a measure line,[6] and the quotient lives in the tensorial quotient of $\mathbb{L}$ and $\mathbb{T}$, another measure line $\frac{\mathbb{L}}{\mathbb{T}}$. The notions of tensorial product and quotient may sound new but no need to worry: they embody just the rules we have been accustomed to since elementary school, e.g., $(6l)(2t) = 12lt$ and $(6l)/(2t) = 3l/t$.[7] A quotient l/t is an example of a velocity value.[8] Similarly, a force value lives in $\mathbb{M}\frac{\mathbb{L}}{\mathbb{T}^2}$, where $\mathbb{T}^2$ abbreviates $\mathbb{TT}$.[9] A tensorial power can be any real number so quantities like $l^{\frac{1}{2}}$, $l^{7.38}$, $l^{\sqrt{2}}$, l^π, $l^{-\pi}$ are meaningful. Negative powers satisfy $\mathbb{L}^{-p} = \mathbb{R}/\mathbb{L}^p$ and $\mathbb{L}^p\mathbb{L}^{-p} = \mathbb{R}$, just as expected since, e.g., for two lengths l_1, l_2, their quotient l_1/l_2 is a real number.[10] For any positive quantity, e.g., a length l, one finds $l^0 = 1$.

Functions like $\sin, \cos, \exp$, and $\ln$ are not meaningful for dimensionful quantities. For example, when we try to extend the definition of $\exp$,

$$\exp x = 1 + x + \frac{1}{2!}x^2 + \frac{1}{3!}x^3 + \cdots, \tag{1}$$

to dimensionful x—e.g., a length l—then each term on the right-hand side lives in different vector spaces: the first one in $\mathbb{R}$, the second in $\mathbb{L}$, the third in $\mathbb{L}^2$ etc. so the sum is meaningless. Only combinations

$$e^{\frac{l}{l_0}}, \qquad \ln\frac{l}{l_0} \qquad etc. \tag{2}$$

are meaningful, where l_0 is also some length value.

---

[4]The typical notation is simply L, T, and M but these letters are overburdened. Also, our typographic choice reflects that, similarly to the famous mathematical sets $\mathbb{R}, \mathbb{C}, \mathbb{Z}, \mathbb{N}, \mathbb{Q}$, the measure lines $\mathbb{L}, \mathbb{T}, \mathbb{M}$ are also famous sets.

[5]The letter $\mathbb{T}$ is already occupied. Temperature is related to heat ($\mathbb{H}$eat).

[6]The tensorial or dyadic product is usually denoted in the way $\mathbb{L} \otimes \mathbb{T}$ or $\mathbb{L} \circ \mathbb{T}$ but usually there is no danger of misunderstanding so we can simply write $\mathbb{LT}$.

[7]Actually, it is just these rules that *define* the tensorial product and quotient of vector spaces.

[8]For velocity *vectors*, and dimensionful vectors in general, the generalization is natural, see the details in the book T. Matolcsi: *Spacetime without reference frames*, Society for the Unity of Science and Technology, Budapest, 2018, ISBN 978-615-80157-3-8 ("Matolcsi_Spacetime_without_Reference_Frames_2018-07-31.pdf" or "Matolcsi_Spacetime_without_Reference_Frames_2018-07-31_two-page_format.pdf" anywhere on the internet).

[9]Mathematics usually writes such a power notation for the Cartesian product of sets, e.g., $\mathbb{R}^2 = \mathbb{R} \times \mathbb{R}$, which contains pairs of values, like $(3.2, 1.84)$. For distinction, we may use $\mathbb{T}^{(2)}$ for $\mathbb{TT}$ but, as long as misunderstanding is unlikely, we will use $\mathbb{T}^2$.

[10]As a special case, $\mathbb{L}^{-1}$ is actually the dual space $\mathbb{L}^*$ of $\mathbb{L}$.

Now comes a notation we are going to find particularly useful. The fact that a length value l lives in measure line $\mathbb{L}$ will be denoted as $\langle\!\langle l \rangle\!\rangle = \mathbb{L}$. Accordingly, with $\langle\!\langle t \rangle\!\rangle = \mathbb{T}$ and $\langle\!\langle m \rangle\!\rangle = \mathbb{M}$, $\left\langle\!\!\left\langle \frac{l}{t} \right\rangle\!\!\right\rangle = \frac{\langle\!\langle l \rangle\!\rangle}{\langle\!\langle t \rangle\!\rangle} = \frac{\mathbb{L}}{\mathbb{T}}$ and $\left\langle\!\!\left\langle m\frac{l}{t^2} \right\rangle\!\!\right\rangle = \langle\!\langle m \rangle\!\rangle \frac{\langle\!\langle l \rangle\!\rangle}{\langle\!\langle t^2 \rangle\!\rangle} = \langle\!\langle m \rangle\!\rangle \frac{\langle\!\langle l \rangle\!\rangle}{\langle\!\langle t \rangle\!\rangle^2} = \mathbb{M}\frac{\mathbb{L}}{\mathbb{T}^2}$.

In vector spaces, one practically (e.g., for numerical purposes) convenient characterization of vectors happens with the aid of a basis. Namely, any vector v of an n dimensional vector space can be uniquely characterized, with the aid of linearly independent vectors v_i ($i = 1, \ldots, n$) by n real numbers c_i that are the coefficients in the expansion $\sum_{i=1}^{n} c_i v_i$. Now, a measure line is a one-dimensional vector space. Correspondingly, only one basis vector is needed to have a basis. For example, any length l can be uniquely expressed, with the aid of a nonzero length l_u as $l = cl_u$. Customarily, a basis vector in a measure line is called a unit of measurement. According to a standard, the notations { }, [ ] are introduced as

$$\{l\} = c, \qquad [l] = l_u. \tag{3}$$

As an example, for a length $l = 3.2\,\text{m}$, $\{l\} = 3.2$ and $[l] = \text{m}$. Unfortunately, other usages of [ ] can also be found in the context of dimensionful quantities, putting the unit into the brackets as $[\text{m}]$, or putting the measure line into the brackets as $[\mathbb{L}]$ (or [L])—sometimes two such different (hence, contradictory) meanings of the bracket are used on the same page of a book.

Confusing, e.g., the number $\{l\} = 3.2$ with l itself—assuming tacitly what the unit, say, $[l] = \text{m}$, is—is a frequent type of mistake that has resulted[11] in death of people (catastrophe of a Korean Air flight, 1999) and in the waste of 300 million USD and of substantial scientific and technological effort (loss of the NASA Mars Climate Orbiter, just in the same year). Everyone should always specify the unit as well; a quantity is meaningless (and dangerous) without the unit.

As a brief repetition:

- Any dimensionful quantity lives in a one-dimensional vector space.

- A unit of measurement is a basis vector in such a one-dimensional vector space.

- The customary rules of product and quotient of dimensionful quantities also apply for the corresponding one-dimensional vector spaces (measure lines).

Units, though hopefully being standardized well enough, are arbitrary in a sense.[12] For a given specific problem, however, there may be some distinguished units, defined by the relevant constants at present. In such a case it is beneficial to use them as units (and to form, from them, units for other measure lines involved) since these embody

---

[11] https://en.wikipedia.org/wiki/Unit_of_measurement#Real-world_implications (as of 2019-08-21).

[12] Why is the second defined as the duration of 9 192 631 770 periods of the radiation corresponding to the transition between the two hyperfine levels of the ground state of the caesium-133 atom, at a temperature of 0K? [Wikipedia: Second (as of 2019-08-21).].

self-scales (characteristic scales or natural scales) for the given situation. In the lucky case, we can make all our quantities dimensionless via these distinguished units.

Let us consider the example of the thermal constitutive relationship of the Van der Waals simple material,

$$p = \frac{RT}{v-b} - \frac{a}{v^2} \tag{4}$$

connecting pressure p, temperature T, and specific volume v, with positive constants a, b, R.

A generally applicable strategy is to identify all the appearing quantities and constants as products of powers of the elementary measure lines $\mathbb{L}$, $\mathbb{T}$, $\mathbb{M}$, and $\mathbb{H}$. For example, the specific volume v lives in $\frac{\mathbb{L}^3}{\mathbb{M}}$,

$$\langle\!\langle v \rangle\!\rangle = \frac{\mathbb{L}^3}{\mathbb{M}}. \tag{5}$$

Then a unit v_{u} is sought in the form

$$v_{\mathrm{u}} = a^{\alpha_v} \cdot b^{\beta_v} \cdot R^{\gamma_v}, \qquad \langle\!\langle v_{\mathrm{u}} \rangle\!\rangle \equiv \frac{\mathbb{L}^3}{\mathbb{M}} = \langle\!\langle a \rangle\!\rangle^{\alpha_v} \langle\!\langle b \rangle\!\rangle^{\beta_v} \langle\!\langle R \rangle\!\rangle^{\gamma_v}, \tag{6}$$

where the powers α_v, β_v, γ_v are determined from the requirement that the total of powers of $\mathbb{L}$ on the right-hand side be 3 (the power of $\mathbb{L}$ on the left-hand side), that the total of powers of $\mathbb{M}$ be -1, the total of powers of $\mathbb{T}$ be 0, and the total of powers of $\mathbb{H}$ be also 0. Hence, we solve a set of linear equations for α_v, β_v, γ_v. The same procedure is to apply for $p_{\mathrm{u}} = a^{\alpha_p} \cdot b^{\beta_p} \cdot R^{\gamma_p}$ and for $T_{\mathrm{u}} = a^{\alpha_T} \cdot b^{\beta_T} \cdot R^{\gamma_T}$.

Yes, this is lengthy and tiring. Therefore, let us now see an alternative route that is much shorter. It is shorter partly because we won't need to identify the $\mathbb{L}$, $\mathbb{T}$, $\mathbb{M}$, and $\mathbb{H}$ content of the measure lines of the constants and of the quantities.

Let us start by recognizing the difference $v - b$ in (4). Such a difference is meaningful only if

$$\langle\!\langle v \rangle\!\rangle = \langle\!\langle b \rangle\!\rangle. \tag{7}$$

This immediately suggests the straightforward choice

$$v_{\mathrm{u}} = b. \tag{8}$$

Next, an analogous consistency requirement tells us that

$$\langle\!\langle p \rangle\!\rangle = \left\langle\!\!\left\langle \frac{a}{v^2} \right\rangle\!\!\right\rangle. \tag{9}$$

We can exploit this as

$$\langle\!\langle p_u \rangle\!\rangle = \langle\!\langle p \rangle\!\rangle = \left\langle\!\!\left\langle \frac{a}{v^2} \right\rangle\!\!\right\rangle = \frac{\langle\!\langle a \rangle\!\rangle}{\langle\!\langle v^2 \rangle\!\rangle} = \frac{\langle\!\langle a \rangle\!\rangle}{\langle\!\langle v \rangle\!\rangle^2} = \frac{\langle\!\langle a \rangle\!\rangle}{\langle\!\langle b \rangle\!\rangle^2}, \tag{10}$$

$$p_u = \frac{a}{b^2}. \tag{11}$$

Finally,

$$\langle\!\langle p \rangle\!\rangle = \left\langle\!\!\left\langle \frac{RT}{v - b} \right\rangle\!\!\right\rangle, \tag{12}$$

$$\frac{\langle\!\langle a \rangle\!\rangle}{\langle\!\langle b \rangle\!\rangle^2} = \left\langle\!\!\left\langle \frac{a}{b^2} \right\rangle\!\!\right\rangle = \langle\!\langle p_u \rangle\!\rangle = \langle\!\langle p \rangle\!\rangle = \frac{\langle\!\langle RT \rangle\!\rangle}{\langle\!\langle v - b \rangle\!\rangle} = \frac{\langle\!\langle R \rangle\!\rangle \langle\!\langle T \rangle\!\rangle}{\langle\!\langle b \rangle\!\rangle} = \frac{\langle\!\langle R \rangle\!\rangle \langle\!\langle T_u \rangle\!\rangle}{\langle\!\langle b \rangle\!\rangle}, \tag{13}$$

$$\langle\!\langle T_u \rangle\!\rangle = \frac{\langle\!\langle a \rangle\!\rangle}{\langle\!\langle b \rangle\!\rangle^2} \frac{\langle\!\langle b \rangle\!\rangle}{\langle\!\langle R \rangle\!\rangle} = \frac{\langle\!\langle a \rangle\!\rangle}{\langle\!\langle b \rangle\!\rangle \langle\!\langle R \rangle\!\rangle}, \qquad T_u = \frac{a}{bR}. \tag{14}$$

Having obtained the distinguished units in either way, we continue with introducing the nondimensionalized quantities as

$$\hat{v} = \frac{v}{v_u}, \qquad \hat{p} = \frac{p}{p_u}, \qquad \hat{T} = \frac{T}{T_u}. \tag{15}$$

In terms of them, the dimensionless counterpart of (4) by replacing v with $\hat{v}v_u$, p with $\hat{p}p_u$, and T with $\hat{T}T_u$ is[13]

$$\hat{p}p_u = \frac{R\hat{T}T_u}{\hat{v}v_u - b} - \frac{a}{(\hat{v}v_u)^2}, \qquad \left| \cdot \frac{1}{p_u} \right. \tag{16}$$

$$\hat{p} = \frac{1}{p_u} RT_u \frac{\hat{T}}{\hat{v}b - b} - \frac{1}{p_u} \frac{a}{v_u^2} \frac{1}{\hat{v}^2} = \frac{b^2}{a} R \frac{a}{bR} \frac{\hat{T}}{b(\hat{v} - 1)} - \frac{b^2}{a} \frac{a}{b^2} \frac{1}{\hat{v}^2}$$

$$= \frac{\hat{T}}{\hat{v} - 1} - \frac{1}{\hat{v}^2}. \tag{17}$$

The result does not contain any dimensionful constant, thus being of a universal form. Accordingly, instead of plotting p–v diagrams for various values of a, b, R, it is enough to plot only a single $\hat{p}$–$\hat{v}$ diagram. Similarly, instead of running a numerical calculation many times for various values of a, b, R, you need only a single run (and then to transform the dimensionless result back to the level of dimensionful quantities).

---

[13] A general suggestion for such transformations: replace the *old* objects *in terms of the new* (and then rearrange), not vice versa: do not try starting with the new ones and attempting to apply the old expressions here or there. And it's not a waste of effort to calculate the inverse transformation: you will anyway need the inverse direction at the end, i.e., when translating anything you obtained at the dimensionless level back to the initial level.

For Van der Waals-like models of simple media—and, according to measurements in real life as well—, there is a distinguished thermodynamical state called the critical point, and then the corresponding critical values v_c, p_c, T_c [which are, for the Van der Waals model, $v_c = 3b$, $p_c = \frac{1}{27}\frac{a}{b^2}$, $T_c = \frac{8}{27}\frac{a}{bR}$] provide distinguished units. The corresponding dimensionless quantities are usually called *reduced quantities*. In thermodynamics, one usually encounters this latter way of nondimensionalizing since it can be performed in any model (that has a critical point) as well as for any real medium (whose critical values are known precisely enough).

As we have seen, all Van der Waals models admit the same nondimensional *thermal* constitutive relationship.[14] When making their *caloric* constitutive relationship (which relates specific internal energy e to T and v)

$$e = \frac{f}{2}RT - \frac{a}{v} \tag{18}$$

nondimensional with these units[15] then the result is

$$\hat{e} = \frac{f}{2}\hat{T} - \frac{1}{\hat{v}}. \tag{19}$$

One can see that no dimensionful parameters have remained in (19), as expected. However, we can also observe that there remained a free dimensionless parameter, f. As a consequence, Van der Waals models with different f prove to be inequivalent in the dimensionless form: the thermal parts are equivalent while the caloric parts not. Accordingly, not states (see Footnote 14) but only some of their quantities may correspond to one another. Different f leads to even qualitatively different behaviour.[16]

Here, one can draw a general moral, actually. Namely, just because you made all your quantities and equations dimensionless, it does not mean that you have

---

[14]This universality has led, in thermodynamics, to the "principle of corresponding states", which can be used for models and real-life media both, and states the expectation that the reduced thermal constitutive functions be the same for different models/media. Among Van der Waals models, this exactly holds, as we have seen. With appropriate care, it can also be used for comparing other models, and for comparing different real-life media, within some reasonable approximation.

[15]In case one uses the critical values v_c, p_c, T_c for nondimensionalization, it is tempting to use $e_c = e(T_c, v_c)$ for making e dimensionless—don't do it. Instead, realize that $\langle\!\langle e \rangle\!\rangle = \langle\!\langle pv \rangle\!\rangle = \langle\!\langle p_c v_c \rangle\!\rangle$ and use $p_c v_c$ as the unit for e. All quantities must be nondimensionalized with respect to the same set of units. Otherwise you can lose consistency. For example, the relationship $\left.\frac{\partial e}{\partial v}\right|_T = T\left.\frac{\partial p}{\partial T}\right|_v - p$ (following from the existence of entropy) can be violated at the dimensionless level. (True story!)

[16]The uncertainty in f has important engineering consequences. Namely, entropy and all other thermodynamical potentials, and thus all related phenomena, depend on this ambiguity. One such phenomenon is whether a medium is 'wet' or 'dry': dry ones are much more advantageous in turbines. Now, a Van der Waals model with $f > 10$ turns out to be dry—see Wikipedia: Working fluid selection (as of 2019-08-21); A. Groniewsky, G. Györke, and A. R. Imre: Description of wet-to-dry transition in model ORC working fluids, *Applied Thermal Engineering* **125** (2017) 963–971; DOI:https://doi.org/10.1016/j.applthermaleng.2017.07.074.

eliminated all free parameters and obtained one universal model.[17] You may have just made the free parameters dimensionless.

Nevertheless, nondimensionalization is useful: we can reduce ourselves to the *essentially free* and *essentially different* parameters.

Finally, some useful formulae: how derivatives can be made dimensionless is illustrated in the following two examples.

$$\left.\frac{\partial p}{\partial v}\right|_T = \left.\frac{\partial (p_u \hat{p})}{\partial v}\right|_T = \left.\frac{\partial (p_u \hat{p})}{\partial \hat{v}}\right|_T \cdot \frac{\mathrm{d}\hat{v}}{\mathrm{d}v} = p_u \left.\frac{\partial \hat{p}}{\partial \hat{v}}\right|_T \cdot \frac{1}{v_u} = \frac{p_u}{v_u} \left.\frac{\partial \hat{p}}{\partial \hat{v}}\right|_T = \frac{p_u}{v_u} \left.\frac{\partial \hat{p}}{\partial \hat{v}}\right|_{\hat{T}},$$

(20)

$$\left.\frac{\partial^2 p}{\partial v^2}\right|_T = \left.\frac{\partial \left(\left.\frac{\partial p}{\partial v}\right|_T\right)}{\partial v}\right|_T = \left.\frac{\partial \left(\frac{p_u}{v_u} \left.\frac{\partial \hat{p}}{\partial \hat{v}}\right|_{\hat{T}}\right)}{\partial v}\right|_T = \left.\frac{\partial \left(\frac{p_u}{v_u} \left.\frac{\partial \hat{p}}{\partial \hat{v}}\right|_{\hat{T}}\right)}{\partial \hat{v}}\right|_T \cdot \frac{\mathrm{d}\hat{v}}{\mathrm{d}v}$$

$$= \frac{p_u}{v_u} \left.\frac{\partial \left(\left.\frac{\partial \hat{p}}{\partial \hat{v}}\right|_{\hat{T}}\right)}{\partial \hat{v}}\right|_{\hat{T}} \cdot \frac{1}{v_u} = \frac{p_u}{v_u^2} \left.\frac{\partial^2 \hat{p}}{\partial \hat{v}^2}\right|_{\hat{T}}. \quad (21)$$

---

[17]Remember these words when doing fluid dynamics.

Printed in the United States
By Bookmasters